1 MONTH OF
FREE
READING

at

www.ForgottenBooks.com

ISBN 978-0-266-91091-6
PIBN 10908466

This book is a reproduction of an important historical work. Forgotten Books uses
state-of-the-art technology to digitally reconstruct the work, preserving the original format
whilst repairing imperfections present in the aged copy. In rare cases, an imperfection in
the original, such as a blemish or missing page, may be replicated in our edition. We do,
however, repair the vast majority of imperfections successfully; any imperfections that
remain are intentionally left to preserve the state of such historical works.

Technical and Bibliographic Notes / Notes techniques et bibliographiques

Institute has attempted to obtain the best original available for filming. Features of this copy which be bibliographically unique, which may alter any of images in the reproduction, or which may ficantly change the usual method of filming are ted below.

Coloured covers /
Couverture de couleur

Covers damaged /
Couverture endommagée

Covers restored and/or laminated /
Couverture restaurée et/ou pelliculée

Cover title missing / Le titre de couverture manque

Coloured maps / Cartes géographiques en couleur

Coloured ink (i.e. other than blue or black) /
Encre de couleur (i.e. autre que bleue ou noire)

Coloured plates and/or illustrations /
Planches et/ou illustrations en couleur

Bound with other material /
Relié avec d'autres documents

Only edition available /
Seule édition disponible

Tight binding may cause shadows or distortion along interior margin / La reliure serrée peut causer de l'ombre ou de la distorsion le long de la marge intérieure.

Blank leaves added during restorations may appear within the text. Whenever possible, these have been omitted from filming / Il se peut que certaines pages blanches ajoutées lors d'une restauration apparaissent dans le texte, mais, lorsque cela était possible, ces pages n'ont pas été filmées.

Additional comments /
Commentaires supplémentaires:

L'Institut a microfilmé le meilleur exemplaire qu' été possible de se procurer. Les détails de cet plaire qui sont peut-être uniques du point de vu ographique, qui peuvent modifier une image repr ou qui peuvent exiger une modification dans la de normale de filmage sont indiqués ci-dessous.

Coloured pages / Pages de couleur

Pages damaged / Pages endommagées

Pages restored and/or laminated. /
Pages restaurées et/ou pelliculées

☑ Pages discoloured, stained or foxed /
Pages décolorées, tachetées ou piquées

Pages detached / Pages détachées

☑ Showthrough / Transparence

☑ Quality of print varies /
Qualité inégale de l'impression

Includes supplementary material /
Comprend du matériel supplémentaire

Pages wholly or partially obscured by errat: tissues, etc., have been refilmed to ensure th possible image / Les pages totaleme partiellement obscurcies par un feuillet d'erra pelure, etc., ont été filmées à nouveau de f; obtenir la meilleure image possible.

Opposing pages with varying colourat discolourations are filmed twice to ensure th possible image / Les pages s'opposant aya colorations variables ou des décoloration filmées deux fois afin d'obtenir la meilleure possible.

opy filmed here has been reproduced thanks
generosity of:

> York University
> Toronto
> Scott Library

ges appearing here are the best quality
le considering the condition and legibility
original copy and in keeping with the
contract specifications.

al copies in printed paper covers are filmed
ning with the front cover and ending on
st page with a printed or illustrated impres-
or the back cover when appropriate. All
original copies are filmed beginning on the
age with a printed or illustrated impres-
and ending on the last page with a printed
strated impression.

st recorded frame on each microfiche
ontain the symbol ➡ (meaning "CON-
ED"), or the symbol ▼ (meaning "END"),
ver applies.

plates, charts, etc., may be filmed at
nt reduction ratios. Those too large to be
ly included in one exposure are filmed
ing in the upper left hand corner, left to
nd top to bottom, as many frames as
ed. The following diagrams illustrate the
d:

L'exemplaire filmé fut reproduit grâce à la
générosité de:

> York University
> Toronto
> Scott Library

Les images suivantes ont été reproduites avec
plus grand soin, compte tenu de la condition
de la netteté de l'exemplaire filmé, et en
conformité avec les conditions du contrat de
filmage.

Les exemplaires originaux dont la couverture
papier est imprimée sont filmés en commen
par le premier plat et en terminant soit par la
dernière page qui comporte une empreinte
d'impression ou d'illustration, soit par le seco
plat, selon le cas. Tous les autres exemplaires
originaux sont filmés en commençant par la
première page qui comporte une empreinte
d'impression ou d'illustration et en terminant
la dernière page qui comporte une telle
empreinte.

Un des symboles suivants apparaîtra sur la
dernière image de chaque microfiche, selon le
cas: le symbole ➡ signifie "A SUIVRE", le
symbole ▼ signifie "FIN".

Les cartes, planches, tableaux, etc., peuvent
filmés à des taux de réduction différents.
Lorsque le document est trop grand pour être
reproduit en un seul cliché, il est filmé à parti
de l'angle supérieur gauche, de gauche à droi
et de haut en bas, en prenant le nombre
d'images nécessaire. Les diagrammes suivan
illustrent la méthode.

| 1 | 2 | 3 |

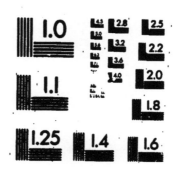

APPLIED IMAGE Inc
1653 East Main Street
Rochester, New York   14609   USA
(716) 482 - 0300 - Phone
(716) 288 - 5989 - Fax

DEPARTMENT OF THE INTERIOR
HONOURABLE CLIFFORD SIFTON, MINISTER

DICTIONARY OF ALTITUDES

IN THE

DOMINION OF CANADA

WITH A RELIEF MAP OF CANADA

BY

JAMES WHITE, F.R.G.S.

GEOGRAPHER

OTTAWA
PRINTED BY S. E. DAWSON, PRINTER TO THE KING'S MOST
EXCELLENT MAJESTY
1903

CONTENTS

CONTENTS

DEPARTMENT OF THE INTERIOR,
OFFICE OF GEOGRAPHER,
OTTAWA, January 3, 1903.

SIR,—I have the honour to transmit herewith my report entitled 'Dictionary of Altitudes in the Dominion of Canada,' accompanied by a relief map of the Dominion, on a scale of 100 miles to one inch. It is supplementary to my 'Altitudes in Canada,' and should have formed Part 2 of that work, but, as its inclusion would have seriously delayed the publication of Part 1, and in view of the demand for such a publication, it was evident that the issuance of Part 1 as a complete work was the lesser evil.

The first work, 'Altitudes in Canada,' is more useful to civil engineers, but the arrangement is not a convenient one for the general public, particularly for anyone wishing to know the altitude, say of a railway station, and ignorant as to the name of the railway line on which it is situated.

It is not presented as absolutely accurate but as the best possible interpretation of the conflicting evidence available at the present time. The figures given in each instance are based on what is considered to be the best authority, but in a few cases where the information at hand did not permit of discrimination, two determinations—with authorities—are given.

The arrangement is alphabetical, by provinces and territories, the locality being given by counties in the ease of the eastern provinces and by latitudes and longitudes in the territories and western provinces where the county system has not been adopted. The latitude and longitude have been taken from the latest maps and charts.

I have the honour to be, sir,

Your obedient servant,

JAMES WHITE
Geographer

JAMES A. SMART, Esq.
Deputy Minister
Department of the Interior

DESCRIPTION OF AUTHORITIES

The following is a descriptive list of authorities quoted in the within report, with short notes as to the methods employed in the determination of the altitudes accredited to them.

ADMIRALTY CHARTS (ADMIR. CHART)

Heights credited to this source were determined by angles measured with a sextant.

BOUNDARY COMMISSION (BDY. COM.)

To this authority are credited all elevations determined by the International Boundary Commissions charged with the determination of the boundary between Canada and the Northern States. All are barometric and between the lake of the Woods and Pacific are of indifferent quality.

BOUNDARY COMMISSION, CANADIAN (BDY. COM. CAN.)

All the elevations given under this authority are in the disputed Alaska-Yukon coast strip and were determined trigonometrically.

CARRUTHERS, ——

Exploration of country between Omineca and Finlay rivers, British Columbia ; determined by aneroid.

CITY ENGINEERS (CITY ENG.)

To this source are credited the elevations of bench marks, city data, etc. in the various cities. They have been measured with a spirit level, and, except on the sea coast where they can be directly referred to sea level, depend upon the altitude of what is considered to be the most accurately determined railway station.

COLEMAN, PROF. A. P.

Exploration in the Rocky mountains between the North Saskatchewan and Athabaska rivers. All determinations are barometrical.

COLLIE, N.

Exploration in the Rocky mountains between the Bow and Athabaska rivers. The elevations were measured by barometer and are, in most instances, too high.

DAWSON, S. J.

Exploration of the country between lakes Superior and Winnipeg. These altitudes were obtained by levelling the principal falls and rapids and estimating the fall in the current and smaller rapids. They have been checked and corrected at several points by railway levels.

vii

DEEP WATERWAYS COMMISSION (D. W. COM.)

The determinations accredited to this source are partly from information supplied by the Canadian Commission and partly from the report of the United States Commission ; the results being derived from the levelling of the Commission and from the work of the United States Lake Survey, with additions and corrections by the commissioners. Since the compilation of 'Altitudes in Canada' the United States Coast Survey has carried out an investigation of the results of precise levelling in the States and has supplemented it with re-levelling and check levelling. The result has been, in general terms, to *lower* the Rouse Point bench mark and the Richelieu and St. Lawrence levels 1·18 feet, and to *raise* lakes Ontario and Erie 0.75, lake Huron 1·14 and lake Superior 1·17 feet relatively to the figures given in the 'Altitudes.' These corrections have been applied to a few city levels directly dependent upon them, to the Great Lakes and to the precise levelling of the Department of Public Works between Rouse Point and Sorel, and between Lachine and Quebec. Pending a general readjustment it has not been considered advisable to make further changes.

GANONG, PROF. W. F.

The elevations credited to this source were measured with the aneroid.

HIND, PROF. H. V.

Assiniboine and Saskatchewan Exploring Expedition, 1858. Elevations were determined barometrically.

GEOLOGICAL SURVEY OF CANADA (GEOL. SURV.)

All the elevations given under this authority are barometric or trigonometric resting on barometric bases.

INTERIOR, DEPARTMENT OF THE

Elevations were determined either by levelling from points on railways or by trigonometric work.

IRRIGATION SURVEY, (IRRIG. SURV.).

The results are from levelling with the spirit level and from trigonometric work.

O'SULLIVAN, H.

Most of these heights were obtained by levelling the falls and rapids in streams and estimating the rise or fall due to the current in the intervening sketches.

PUBLIC WORKS, DEPARTMENT OF (PUB. WORKS)

These elevations are the results of exact levels.

RAILWAYS AND CANALS, DEPARTMENT OF (RYS. & CANALS)

All elevations are from barometric determinations.

UPHAM, W.

These results are from barometric determinations.

WILCOX, W. D.

Elevations were determined barometrically and by vertical angles.

NAMES OF RAILWAYS

A. C. & H. B. R ...... Algoma Central and Hudson Bay Railway
A. & L. S. R ...... Atlantic and Lake Superior Railway
A. R. & C. Co ...... Alberta Railway and Coal Company's Railway
A. S. R ...... Albert Southern Railway
B. M. & A. R ...... Bruce Mines and Algoma Railway
B. & M. R. ...... Buctouche and Moncton Railway (New Brunswick)
B. & M. R. ...... Boston and Maine Railroad (Quebec)
B. of Q. R ...... Bay of Quinté Railway
B. & W. R ...... Brockville, Westport and Sault Ste. Marie Railway
Car. R ...... Caraquet Railway
C. A. R ...... Canada Atlantic Railway
C. C. & R. Co ...... Canada Coals and Railway Company
C. C. Co. ...... Canadian Copper Company
C. E. R ...... Canada Eastern Railway
C. M. T. R. ...... Chignecto Marine Transport Railway
C. N. R ...... Canadian Northern Railway
C. O. R. ...... Central Ontario Railway
C. P. R. ...... Canadian Pacific Railway
C. R. of N. B. ...... Central Railway of New Brunswick
C. R. of N. S ...... Central Railway of Nova Scotia
C. R. & C. Co ...... Cumberland Railway and Coal Company
D. A. R. ...... Dominion Atlantic Railway
E. & H. R. ...... Elgin and Havelock Railway
E. & N. R ...... Esquimalt and Nanaimo Railway
G. N. R. ...... Great Northern Railway
G. S. R ...... Gulf Shore Railway
G. T. R ...... Grand Trunk Railway
H. & Y. R. ...... Halifax and Yarmouth Railway
I. B. & O. R. ...... Irondale, Bancroft and Ottawa Railway
I. C. R ...... Intercolonial Railway
I. & R. R. ...... Inverness and Richmond Railway
K. & P. R. ...... Kingston and Pembroke Railway
K. & S. R . ... Kaslo and Slocan Railway
L. E. & D. R. R. ...... Lake Erie and Detroit River Railway
L. & M. R. ...... Lotbinière and Megantic Railway
M. & B. R. V. R ...... Missisquoi and Black River Valley Railway
M. C. R. ...... Michigan Central Railroad
Me. C. R ...... Maine Central Railroad
M. & G. C. R. ...... Montfort and Gatineau Colonization Railway
M. & N. S. R. ...... Manitoulin and North Shore Railway
M. R. ...... Midland Railway
N. B. & P. E. I. R ...... New Brunswick and Prince Edward Island Railway
N. & F. S. R ...... Nelson and Fort Sheppard Railway
N. S. S. R. ...... Nova Scotia Southern Railway
N. S. S. C. R ...... Nova Scotia Steel Company's Railway
N. Y. C. R. ...... New York Central Railroad, Adirondack Division
N. Y. & O. R. ...... New York and Ottawa Railway
O. B. & N. R ...... Ontario, Belmont and Northern Railway
O. M. R. ...... Orford Mountain Railway
O. N. & W. R ...... Ottawa, Northern and Western Railway
P. E. I. R. ...... Prince Edward Island Railway
P. P. J. R. ...... Pontiac Pacific Junction Railway

DEPARTMENT OF THE INTERIOR

Q. C. R. .. .. Quebec Central Railway
Q. & L. St. J. R.. ....Quebec and Lake St. John Railway
Q. R. L. & P. Co ... .Quebec Railway, Light and Power Company
Q. S R ... .Quebec Southern Railway
R. M. R... ....Red Mountain Railway
R. & W. R...... . ...Restigouche and Western Railway
St. C. & N. C. h................. ....St. Catharines and Niagara Central Railway
St. J. V. & R. L. R...... ....St. John Valley and River du Loup Railway
St. L. & R. R..... ....St. Louis and Richibucto Railway
St. M. R. R.... .... St. Mary River Railway
S. L. R............... . ...Shore Line Railway
S. & L. R........... ....Sydney and Louisburg Railway
S. S. R..... . ...South Shore Railway
Tem. R........ ....Temiscouata Railway
T. H. & B. R... .... .Toronto, Hamilton and Buffalo Railway
T. I. R......... .... Thousand Islands Railway
T. L. E. & P. R ....Tilsonburg, Lake Erie and Pacific Railway
V. & I. I. R.... .... Vancouver and Lulu Island Railway
V. & S. R....................... . .. Victoria and Sydney Railway
W. P. & Y. R ....White Pass and Yukon Railway

DICTIONARY OF ALTITUDES

IN THE

DOMINION OF CANADA

ALBERTA

Locality	Elev.	Lat.		Long.		Authority
		°	'	°	'	
Aberdeen, Mount..	10,480	51	23	116	15	Interior
Agnes, Lake..	6,850	51	25	116	14	Wilcox
Airdrie..	3,538	51	17	114	01	C. P. R.
Alberta, Mount..	13,500	52	14	117	36	Collie
Allsmoke hill..	6,894	50	45	114	41	Irrig. Surv.
Anthracite..	4,490	51	11	115	30	C. P. R.
Arcs, Lac des..	4,217	51	03	115	11	C. P. R.
Assiniboine, Mount.................................	11,860	50	56	115	42	Interior
Athabaska, Mount.................................	11,900	52	07	117	11	Collie
Athabaska pass....................................	6,925	52	27	118	20	C. P. R.
do	5,710					Coleman
Aylmer, Mount....................................	10,365	51	20	115	26	Interior
Balfour, Mount...................................	10,330	51	34	116	28	Interior
Ball, Mount.......................................	10,930	51	09	116	00	Interior
Banff..		51	11	115	35	
Station...	4,521					C. P. R.
Summit, 2·5 miles E..........................	4,573					C. P. R.
Forty-mile brook, water, 4,512; rail........	4,518					C. P. R.
Bantry	2,471	50	32	111	50	C. P. R.
Barwell hill	6,249	50	47	114	39	Irrig. Surv.
Bassano...	2,589	50	47	112	28	C. P. R.
Battle lake..	2,795	52	57	114	10	Geol. Surv.
Bear lake..	2,648	52	57	113	37	Geol. Surv.
Beaver hills.......................................	2,525	53	30	113	00	Geol. Surv.
Beaver lake.......................................	2,173	53	30	112	30	Geol. Surv.
Beddington..	3,464	51	09	114	02	C. P. R.
Beehive mountain..............................	7,380	51	25	116	14	Interior
Beehive mountain..............................	8,500	50	04	114	44	Geol. Surv.
Bench Marks—						Irrig. Surv.
*N.E. cor., Tp. 3, R. XVIII, W. 4th Mer..	3,468·9	49	16	112	17	
Do 4, R. XVII do	3,243·0	49	21	112	09	
Do 3, R. XVII do	3,253·1	49	16	112	09	
S.E. cor., Tp. 3, R. XVII do	3,801·7	49	10	112	17	
Do 3, R. XVII do	3,580·6	49	10	112	09	
Do 3, R. XIX do	3,804·2	49	10	112	25	
Do 3, R. XVI do	3,472·6	49	10	112	01	
At junction of N. and S. branches Milk river in Tp. 2, R. XVIII, W. of 4th Mer..............	3,543·5	49	08	112	22	
At intake of proposed Milk river irrigation canal in S.E. ¼, Sec. 29, Tp. 2, R. XVII, W. of 4th Mer.	3,484·0	49	09	112	14	
N.E. cor., Tp. 5, R. XVII, W. of 4th Mer....	3,113·9	49	26	112	09	
Do 5, R. XVIII do	3,163·0	49	26	112	17	
Do 6, R. XVIII do	3,132·8	49	31	112	17	

* When not otherwise stated these benches are iron bars driven to within about 10 inches of the surface

ALBERTA

Locality	Elev.	Lat.		Long.		Authority
		°	′	°	′	
Bench Marks—						Irrig. Surv.
N.E. cor., Tp. 4, R. XVIII, W. 4th Mer....	3,203·8	49	21	112	18	
Do 4, R. XIX do	3,385·6	49	21	112	26	
Do 4, R. XX do	3,435·4	49	21	112	34	
Do 3, R. XXI do	4,211·5	49	16	112	42	
Do 5, R. XXI do	3,211·1	49	26	112	42	
Do 7, R. XXI do	3,020·8	49	37	112	43	
N.E. cor., Tp. 7, R. XVII, W. 4th Mer...........	2,985·7	49	37	112	10	
Do 6, R. XVI do	3,077·8	49	31	112	02	
Do 7, R. XIX do	3,058·2	49	37	112	26	
Do 9, R. XVIII do	2,692·9	49	47	112	18	
Do 9, R. XX do	2,808·0	49	47	112	34	
N.E. cor., Sec. 35, Tp. 8, R. XXII, W. 4th Mer....	2,700·8	49	42	112	52	
¼ sec. cor., on N. boundary of Sec. 33, Tp. 8, R. XXI, W. 4th Mer......	3,002·4	49	42	112	47	
N.E. cor., Sec. 33, Tp. 9, R. XXI, W. 4th Mer....	2,931·1	49	47	112	46	
Do 33, Tp. 3, R. XXII do	3,717·3	49	16	112	50	
On ¼ sec. mound, E. bdy., Sec. 2, Tp. 4, R. XXII, W. 4th Mer......	3,612·4	49	16	112	51	
On S. boundary of Tp. (215 ft. E. of N. branch of Milk R.) Tp. 1, R. XXIII, W. 4th Mer....	4,155·4	49	00	113	00	
S.E. cor., Tp. 1, R. XXIII, W. 4th Mer.	4,619·1	49	00	112	57	
N.E. cor. Sec. 36, Tp. 4, R. XXIII, W. 4th Mer.....	3,431·8	49	21	112	58	
Do 34, Tp. 4, R. XXIV do	3,624·0	49	21	113	08	
41 chains N. of N.E. cor., Sec. 24, Tp. 6, R. XXIII, W. 4th Mer......	3,022·9	49	30	112	58	
40 chains S. of N.E. cor., Sec. 25, Tp. 8, R. XXIII, W. 4th Mer......	2,789·2	49	40	112	59	
¼ sec. cor. on N. boundary of Sec. 33, Tp. 8, R. XXIV, W. 4th Mer......	2,980·2	49	42	113	11	
N.E. cor., Tp. 1, R. XXIV, W. 4th Mer..........	4,092·7	49	05	113	05	
Do 2, R. XXIV do	3,948·5	49	10	113	05	
Do 1, R. XXV do	3,961·8	49	05	113	13	
On S.E. ¼ Sec. 36, Tp. 1, R. XXV, W. 4th Mer. at in-take of St. Mary irrigation canal....	3,877·8	49	04	113	13	
S.E. cor., Sec. 17, Tp. 2, R. XXIV, W. 4th Mer....	3,847·7	49	08	113	10	
Do 6, Tp. 1, R. XXV do	4,160·5	49	00	113	19	
25 chs. S. of N.E. cor., Sec. 36, Tp. 2, R. XXVI, W. 4th Mer......	3,817·1	49	10	113	20	
N.E. cor., Tp. 3, R. XXIV, W. 4th Mer..	3,702·3	49	16	113	06	
Near centre of pits marking trail survey at W. end of bridge over St. Mary river in Sec. 11, Tp. 3, R. XXV, W. 4th Mer....	3,639·9	49	13	113	16	
N.E. cor., Tp. 5, R. XXVI, W. 4th Mer...........	3,362·4	49	26	113	22	
Do 4, R. XXVII do	3,519·7	49	21	113	30	
S.E. cor., Sec. 1, Tp. 3, R. XXVIII, W. 4th Mer.	4,162·1	49	10	113	38	
Do 1, Tp. 3, R. XXIX do	4,376·8	49	10	113	45	
21 chains N. of N.E. cor., Tp. 1, R. XXVIII, W. 4th Mer., on shore of lake......	4,582·3	49	05	113	37	
N.E. cor., Tp. 1, R. XIX, W. 4th Mer	4,894·7	49	05	113	44	
E. side of trail, between Waterton lakes, in Sec. 25, Tp. 1, R. XXX, W. 4th Mer	4,203·3	49	04	113	53	
N.E. cor., Sec. 12, Tp. 9, R. XXVI, W. 4th Mer ...	3,084·2	49	41	113	23	
Do 34, Tp. 8, R. XXVI do ...	3,124·5	49	42	113	26	
39 chains W. of N.E. cor., Sec. 33, Tp. 8, R. XXVI, W. 4th Mer......	3,181·8	49	42	113	28	
N.E. cor. Tp. 9, R. 27, W. 4th Mer......	3,163·5	49	47	113	31	
Station 52, Macleod and Pincher Creek trail, Sec. 7, Tp. 7, R. XXVIII, W. 4th Mer	3,379·7	49	33	113	47	
Station 55, Macleod and Pincher Creek trail, Sec. 12, Tp. 7, R. XXIX, W. 4th Mer....	3,383·4	49	33	113	48	
N.E. cor., Tp. 7, R. I, W. 5th Mer......	3,777·9	49	37	114	00	
Do 6, R. XXX, W. 4th Mer	3,643·2	49	31	113	54	
Do 6, R. I, W. 5th Mer	3,878·9	49	31	114	00	
¼ sec. mound on E. bdy., Sec. 12, Tp. 6, R. I., W. 5th Mer	4,055·8	49	27	114	00	
N.E. cor., Tp. 5, R. XXIX, W. 4th Mer......	3,754·1	49	26	113	46	
Do Tp. 10 R. XXIII, W. 4th Mer	3,191·8	49	52	112	59	
Do 12, R. XXIII do	3,261·2	50	03	113	00	

ALBERTA

Locality	Elev.	Lat.		Long.		Authority
		′	″			
Bench Marks--						Irrig. Surv.
N.E. cor., Tp. 12, R. XXV, W. 4th Mer.....	3,180·1			113	16	
Do 12, R. XXVI do	3,220·8	50	03	113	24	
Do 12, R. XXVIII do	3,362·7	50	03	113	41	
5 enaims S. of N.E. cor., Sec. 12, Tp. 15, R. XXVI, W. 4th Mer......	3,163·6	50	15	113	26	
1·43 chains N. of N.E. cor., Sec. 25, Tp. 15, R. XXVIII, W. 4th Mer........	3,368·6		18	113	45	
10·50 chains S. of N.E. cor., Sec. 12, Tp. 16, R. XXVIII, W. 4th Mer..........	3,301·0		20	113	43	
Near north end of Can. Pac. Ry. bridge over S. Br. Mosquito Ck. (10 ft. S. of first telegraph post)....	3,341·2		21	113	47	
N.E. cor., Tp. 16, R. XXVI, W. 4th Mer..	3,389·1		24	113	26	
20 chs. W. of N.E. cor., Sec. 31, Tp. 16, R. XXVI, W. 4th Mer....	3,198·1	50 50	24	113	33	
5 chs. E. of N.E. cor., Sec. 31, Tp. 16, R. XXVIII, W. 4th Mer........	3,366·2	50	24	113	49	
12 chs. W. of N.E. cor., Sec. 35, Tp. 16, R. XXIX, W. 4th Mer........	3,376·6	50	24	113	52	
N.E. cor., Sec. 36, Tp. 16, R. II, W. 5th Mer..	4,427·5	50	24	114	08	
20 chs. W. of N.E cor., Sec. 33, Tp. 16, R. II, W. 5th Mer........	3,998·6	50	24	114	13	
55 chs. S. of N.E. cor., Sec. 1, Tp. 18, R. II, W. 5th Mer.	3,718·6		29	114	08	
47 chs. S. of N.E. cor., Sec. 17, R. II, W. 5th Mer..	3,756·7		28	114	08	
S.E. cor., Sec. 1, Tp. 19, R. XXVIII, W. 4th Mer..	3,291·8		34	113	45	
N.E. cor., Tp. 19, R. XIX, W. 4th Mer...........	3,422·6		40	113	53	
Do 20, R. XXVIII do	3,564·7		45	113	44	
Do 21, R. XXVIII do	3,328·4	50 50	50	113	44	
10 chs. E. of N.E. cor., Sec. 31, Tp. 16, R. XX, W. 4th Mer.......	3,237·7	50	45	113	51	
North boundary, Sec. 35, Tp. 20, R. I, W. 5th Mer., at Can. Pac. Ry. crossing.....	3,497·5	50	45	114	02	
Sec. 29, Tp. 20, R. XXIX, W. 4th Mer., at highway bridge over Sheep river......	3,444·0	50	43	113	59	
N.E. cor., Sec. 36, Tp. 20, R. II, W. 5th Mer..	3,823·7	50	45	114	08	
49 chs. S. of N.E. cor., Sec. 25, Tp. 20, R. II, W. 5th Mer.	3,653·8	50	43	114	08	
25 chs. S. of N.E. cor., Sec. 1, Tp. 22, R. II, W. 5th Mer.	3,562·4	50	51	114	08	
20 ft. S. of N.E. cor., Sec. 2, Tp. 21, R. III, W. 5th Mer.	3,841·1	50	46	114	18	
3 ft. N. of S. W. cor. of school plot in S.E. cor., Sec. 22, Tp. 22, R. III. W. of 5th Mer....	3,792·0	50	53	114	19	
N.E. cor., Sec. 4, Tp. 22, R. III, W. 5th Mer.....	3,914·6	50	51	114	21	
Do 24, Tp. 22, R. IV do	3,989·2	50	54	114	25	
Do 36, Tp. 22, R. IV do	4,538·6	50	55	114	25	
E. bdy. Sec. 4, Tp. 23, R. I, W. 5th Mer., about 75 ft. W. of highway bridge over Fish creek.	3,389·0	50	56	114	04	
35 chs. S. of N.E. cor., Sec. 1, Tp. 23, R. II, W. 5th Mer.	3,512·1	50	56	114	08	
At intersection of E. bdy., Tp. 23, R. XXIV, W. 4th Mer. with Can. Pac. Ry. near Namaka station..	2,948·0	50	55	113	13	
N.E. cor., Tp. 24, R. XXIV, W. 4th Mer.....	3,059·0	51	00	113	13	
Do 23, R. XXIV do	3,111·5	51	00	113	21	
S.W. cor., Tp. 23, R. XXV do	3,256·2	50	55	113	30	
N.E. cor., Sec. 36, Tp. 23, R. XXVI, W. 4th Mer...	3,237·6	51	00	113	30	
Do 36, Tp. 24 do do ...!	3,070·1	51	06	113	30	
Do 36, Tp. 25 do do ...!	3,075·6	51	11	113	30	
Do 36, Tp. 25, R. XXVII do ...!	3,113·1	51	11	113	38	
E. bdy., Sec. 13, Tp. 23, R. XXVIII, W. 4th Mer. at Can. Pac. Ry. crossing.	3,354·1	50	58	113	46	
N.E. cor., Tp. 24, R. XXVIII W. 4th Mer........	3,301·7	51	06	113	46	
Do. 25, R. XXVIII, do	3,357·9	51	11	113	46	
E. bdy., Sec. 13, Tp. 23, R. XXIX, W. 4th Mer. at S.E. corner, Shepard station ground	3,362·5	50	57	113	55	
N.E. cor., Tp. 24, R. XXIX, W. 4th Mer....	3,536·4	51	06	113	55	
Do Sec. 10, Tp. 23, R. XXIX, W. 4th Mer...	3,375·3	51	02	113	57	

ALBERTA

Locality	Elev.	Lat.		Long.		Authority
		° ′		′ ″		
Bench Marks—						Irrig. Surv.
N.E. cor., Tp. 24, R. II, W. 5th Mer..............	3,631·6	51	06	114	08	
Nose hill, sandstone rock, 20 chs. E. of N.W. cor. Sec. 31, Tp. 24, R. I., W. 5th Mer..	3,865·4	51	06	114	08	
52 chs. S. of N.E. cor., Sec. 25, Tp. 23, R. II, W. 5th Mer..................	3,514·0	50	59	114	08	
10 chs. W. of N.E. cor., Sec. 34, Tp. 24, R. II, W. 5th Mer...........	3,492·4	51	06	114	11	
N.E. cor., Tp. 24, R. III, W. 5th Mer...........	3,808·1	51	06	114	17	
¼ sec. cor. on E. bdy. Sec. 7, Tp. 24, R. II, W. 5th Mer..................	3,601·4	51	01	114	15	
N.E. cor., Sec. 34, Tp. 24, R. I, W. 5th Mer.....	3,429·4	51	06	114	03	
1 chain N.W. of ¼ sec. cor. on E. bdy., Sec. 13, Tp. 24, R. IV, W. 5th Mer.............	3,807·4	51	03	114	25	
N.E. cor., Tp. 24, R. IV, W. 5th Mer...........	3,983·8	51	06	114	25	
35 chs. W. of N.E. cor., Sec. 31, Tp. 24, R. IV, W. 5th Mer...........	3,986·5	51	06	114	32	
¼ Sec. md. on E. boundary, Sec. 25, Tp. 25, R. IV, W. 5th Mer...........	3,664·3	51	09	114	25	
N.E. cor., Tp. 26, R. IV, W. 5th Mer	4,238·8	51	16	114	25	
Do 26, R. XXVII, W. 4th Mer........	3,062·8	51	16	113	38	
Do 27, R. XXVII, do	3,053·8	51	21	113	40	
Do 28, R. XXVII do	3,137·2	51	26	113	40	
Do 28, R. XXVIII do	3,190·0	51	26	113	48	
Do 29, R. XXVIII do	3,241·7	51	31	113	48	
Near N.E. cor., Sec. 33, Tp. 28, R. XXVIII, W. 4th Mer.....	3,161·0	51	26	113	52	
N. bdy., Sec. 35, Tp. 28, R. I, W. 5th Mer. at crossing of Edmonton branch, Can. Pac. Ry......	3,586·6	51	26	114	02	
N.E. cor., Tp. 26, R. II, W. 5th Mer..............	3,895·2	51	16	114	08	
Do 28, R. II do	3,768·6	51	26	114	08	
Ledge of sandstone, about 6 chs. W. of N.E. cor., Tp. 28, R. III, W. 5th Mer.....	3,632·5	51	26	114	17	
N.E. cor., Sec. 28, R. IV, W. 5th Mer...........	3,945·2	51	26	114	25	
¼ sec. mound N. bdy., Sec. 33, Tp. 28, R. IV, W. 5th Mer.....	3,863·0	51	26	114	30	
N.E. cor., Tp. 29, R. XXIX, W. 4th Mer........	3,144·8	51	31	113	56	
Do 29, R. I, W. 5th Mer	271·1	51	31	114	00	
Do 30, R. I do	3,971·6	51	37	114	00	
Do 31, R. I do	51·0	51	42	114	00	
Do 31, R. II do	3,366·4	51	42	114	09	
Do 32, R. II do	3,345·6	51	48	114	09	
Do 29, R. III do	3,539·8	51	31	114	17	
Do 30, R. III do	3,560·3	51	37	114	17	
Do 31, R. III do	3,529·8	51	42	114	17	
Do 32, R. III do	3,424·8	51	48	114	17	
¼ sec. mound, E. bdy. Sec. 10, Tp. 33, R. III, W. 5th Mer.....	3,376·2	51	49	114	20	
N.E. cor., Tp. 29, R. IV, W. 5th Mer..........	3,852·5	51	31	114	25	
Do 30, R. IV do	3,456·9	51	37	114	25	
Do 31, R. IV do	3,608·4	51	42	114	25	
3 chs. E. of N.E. cor., Tp. 32, R. IV, W. 5th Mer...	3,540·2	51	48	114	25	
N.E. cor., Tp. 32, R. V, W. 5th Mer. . ..	3,679·0	51	48	114	34	
¼ sec. mound, N.E. Sec. 34, Tp. 32, R. V, W. 5th Mer.....	3,560·0	51	48	114	38	
Bert hill..........................	5,077	50	51	114	32	Irrig. Surv.
Biche, Lac la......................	1,650	54	55	112	00	Geol. Surv.
Big hill...........................	4,275	51	10	114	27	Geol. Surv.
Birch lake........................	2,107	53	20	111	37	Geol. Surv.
Black butte.......................	2,190	53	42	112	00	Geol. Surv.
Blackfalds........................	2,876	52	23	113	48	C. P. R.
Black Spring ridge................	3,550	50	00	112	55	Geol. Surv.
Blairmore.........................	4,226	49	36	114	27	
Station.......	4,226					C. P. R.
Crownest river, 1·3 mile E., water 4,172 ; rail......	4,182					C. P. R.
Bonnet peak........................	10,200	51	25	115	53	Interior
Bourgeau mountain................	9,517	51	08	115	46	Interior
Bow lake, Lower..................	5,530	51	35	116	22	Geol. Surv.
Bow lake, Upper..................	6,200	51	38	116	25	Geol. Surv.
Bowden............................	3,232	51	55	114	01	C. P. R.

ALBERTA

Locality	Elev.	Lat.	Long.	Authority
Brazeau lake.......................	6,300	52 29	117 08	Coleman
Brocket........		49 33	113 46	
Station................................	3,500			C. P. R.
Summit, 0·8 mile W...................	3,529			C. P. R.
Pincher creek, bed 3,383; rail..................	3,493			C. P. R.
Brown, Mount...	9,050	52 27	118 22	Coleman
Brunton........................	3,308	49 17	112 12	A. R. & C. Co.
Bryce, Mount..................	13,000	51 58	117 20	Collie
Buck lake.................	2,970	53 00	114 45	Geol. Surv.
Buffalo hills.................	3,857	50 35	113 13	Geol. Surv.
Buffalo lake..............	2,536	52 30	112 55	Geol. Surv.
Bullrush mountain	7,200	53 18	117 55	Geol. Surv.
Burdett.......................	2,577	49 50	111 30	C. P. R.
Burmis (formerly Livingstone)...................		49 35	114 16	
Station ...	3,995			C. P. R.
Crowsnest river, 4·2 miles E., water, 3,795; rail.....	3,815			C. P. R.
Calgary.....................	3,428	51 02	114 03	
Station.............................				C. P. R.
B.M., at in-take of Bow River canal...........	3,388·4			Irrig. Surv.
B.M., at Can. Pac. Ry. bridge over Elbow river.....	3,403·1			Irrig. Surv.
Rail level, Can. Pac. Ry. opposite N.W.N.P. barracks	3,411·0			Irrig. Surv.
B.M., N.W. corner of post office.....	3,425·0			Irrig. Surv.
B.M., S.E. corner of court house	3,430·0			Irrig. Surv.
B.M., south side of chimney of water works pump house	3,434·0			Irrig. Surv.
Bow river, 3·3 miles E., water, 3,354; rail.	3,373			C. P. R.
Elbow river, 0·9 mile E., high water, 3,401; low water, 3,394; rail..	3,410			C. P. R.
Calgary junction.....................	3,410	51 02	114 01	C. P. R.
Calgary and Edmonton junction..........	3,139	49 43	113 27	C. P. R.
Canmore		51 05	115 21	
Station...........	4,284			C. P. R.
Bow river, 5·8 miles W., high water, 4,352; low water	4,349			C. P. R.
Carstairs	3,464	51 34	114 06	C. P. R.
Cascade..	4,337	51 10	115 43	C. P. R.
Cascade mountain............	9,826	51 14	115 34	Interior
Cassils	2,517	50 34	112 02	C. P. R.
Castle Mountain station	4,660	51 16	115 54	C. P. R.
Castle mountain..........................	9,030	51 18	115 56	Interior
Cayley.....		50 27	113 51	
Station............................	3,503			C. P. R.
Summit, 3·9 miles N.................	3,521			C. P. R.
Cheadle.........................	3,189	51 02	113 30	C. P. R.
Chin....	2,783	49 45	112 26	C. P. R.
Chiniquy hill	5,287	51 06	114 52	Irrig. Surv.
Claresholm....	3,530	50 02	113 35	C. P. R.
Cluny	2,850	50 50	112 52	C. P. R.
Coaldale......................	2,826	49 43	112 36	C. P. R.
Cochrane................................	3,749	51 11	114 28	C. P. R.
Coleman, Mount....	11,000	52 06	116 55	Collie
Collie, Mount......	10,500	51 35	116 35	Outram
Columbia, Mount, about............	*14,000	52 05	117 32	Collie
Cooking lake..............................	2,400	53 28	113 00	Geol. Surv.
Copper mountain.....	9,160	51 13	115 53	Interior
Coutts.........................		49 00	111 57	
Station	3,463			A. R. & C. Co.
Summit, 1·5 mile N	3,493			A. R. & C. Co.
Cowley		49 34	114 04	
Station...........................	3,831			C. P. R.
South fork Oldman river, 2·5 miles E., water 3,584; rail..	3,717			C. P. R.
Coxcomb hill.................................	7,169	51 00	114 54	Irrig. Surv.
Crossfield.......................		51 25	114 02	
Station..	3,622			C. P. R.
Summit, 0·5 mile N...........	3,627			C. P. R.
Crowfoot....	2,698	50 49	112 39	C. P. R.
Crowsnest junction........	2,997	49 42	112 47	C. P. R.
Crowsnest lake....	4,409	49 38	114 38	C. P. R.
Crowsnest mountain....	7,800	49 40	114 36	Geol. Surv.

* Probably overestimated

ALBERTA

Locality	Elev.	Lat.		Long.		Authority
Crowsnest pass, railway summit..	4,449	49	38	114	42	C. P. R.
Crowsnest pass, trail summit	4,845	49	33	114	44	Geol. Surv.
Dalhousie, Mount...........................	8,000	52	30	116	45	Geol. Surv.
Daly, Mount	10,255	51	31	116	23	Interior
Deltaform, Mount...........................	10,945	51	18	116	15	Interior
De Winton....		50	49	1.4	01	
Station.... ...	3,621					C. P. R.
Summit 0·6 mile S..	3,622					C. P. R.
Devils or Minnewanka lake................	4,800	51	15	115	25	Interior
Devils Head mountain	9,205	51	21	115	16	Interior
Devils Pine lake........................	2,910	52	06	113	26	Geol. Surv.
Diadem peak.........................	11,500	52	16	117	31	Collie
Didsbury................................		50	39	114	08	
Station. ...	3,300					C. P. R.
Rosebud river, 3·0 miles N., bed 3,290; rail..	3,298					C. P. R.
Dome mountain......	11,850	52	06	117	19	Collie
Douglas, Mount.........	8,000	51	32	115	50	Geol. Surv.
Douglas peak...........................	11,700	52	08	117	21	Collie
Dowling lake........................ ...	2,563	51	45	112	00	Geol. Surv.
Driedmeat lake.. ...	2,255	52	50	112	45	Geol. Surv.
Dutchman hill...........................	4,964	50	39	114	25	Irrig. Surv.
Duthil (closed).........................	4,387	51	10	115	28	C. P. R.
Eagle lake	3,006	51	00	113	20	Irrig. Surv.
Edith, Mount.	8,370	51	12	115	30	Interior
Edmonton		53	33	113	30	
Station......	2,188					C. P. R.
North Saskatchewan river, high water, 2,012; water, Sept. 20, 1890...	1,995					C. P. R.
Top of bank, N. side of valley..............	2,177					C. P. R.
Egg lake....................	2,045	53	50	112	15	Geol. Surv.
Egg lake...................	2,094	51	30	112	07	Geol. Surv.
Eldon	4,817	51	23	116	02	C. P. R.
Elk mountains.......................	9,000	50	30	114	45	Geol. Surv.
End mountain	7,856	51	12	115	08	Interior
Etzikom coulée, in Tp. 6, R. XVIII......	2,974	49	30	112	20	Irrig. Surv.
Fairholme mountains, north	9,305	51	09	115	20	Interior
Fairholme mountains, south............	9,205	51	07	115	19	Interior
Fairview mountain	8,875	51	24	116	13	Interior
Fatigue mountain	9,697	51	02	115	41	Irrig. Surv.
Fifteen-mile Butte station...........	3,125	49	28	112	42	St. M. R. R.
Fish butte........	4,009	50	55	114	31	Irrig. Surv.
Flagstaff hill......................	2,500	52	32	111	33	Irrig. Surv.
Forbes, Mount.......................	13,400	51	48	116	56	Collie
Forget-me-not mountain.............	7,577	50	45	114	47	Irrig. Surv.
Forks hill	5,764	50	39	114	38	Irrig. Surv.
Freshfield, Mount..	12,000	51	40	116	56	Collie
Geikie, Mount	11,000	52	43	118	22	Geol. Surv.
Gleason hill......	5,919	50	42	114	35	Irrig. Surv.
Gleichen........................	2,952	50	52	113	03	C. P. R.
Glenbow..........................	3,620	51	09	114	26	C. P. R.
Goat range......... .	9,290	51	01	115	27	Interior
Gordon, Mount	11,130	51	36	116	31	Interior
Gould Dome	10,125	49	56	114	39	Geol. Surv.
Grassy Lake station.	2,652	49	49	111	41	C. P. R.
Grotto mountain...	8,870	51	05	115	16	Interior
Gull lake.	2,905	52	30	114	00	Geol. Surv.
Hand hills........................	3,575	51	30	112	20	Geol. Surv.
Hastings lake	2,380	53	26	112	55	Geol. Surv.
Hector, Mount	11,235	51	34	116	15	Interior
High River........		50	34	113	53	
Station.......	3,394					C. P. R.
Highwood river, high water, 3,386; low water..	3,380					C. P. R.
B.M., S.E. corner stone, 5th course, High River Trading Co's store	3,395·2					Irrig. Surv.
Hobbema.. ..	2,612	52	51	113	27	C. P. R.
Hoffmann mountain........................	6,572	50	36	114	41	Irrig. Surv.
Hole-in-the-wall mountain.	9,184	51	12	115	42	Interior
Hooker, Mount......	10,505	52	26	118	18	Wilcox

ALBERTA

Locality	Elev.	Lat.	Long.	Authority
Howse pass............................	4,500	51 46	116 45	C. P. R.
Howse peak............................	10,000	51 49	116 40	Collie
Hungabee, Mount......................	11,305	51 20	116 16	Interior
Inglismaldie, Mount..................	9,715	51 15	115 25	Interior
Innisfail............................	3,087	52 02	113 57	C. P. R.
Island Lake, extreme high water, 4,427; extreme low water.............	4,420	49 38	114 43	C. P. R.
Jacques, Roche.......................	8,500	53 04	117 58	Geol. Surv.
Jasper lake..........................	3,256	53 06	118 00	C. P. R.
Junction mountain....................	6,720	50 35	114 43	Irrig. Surv.
Kananaskis...........................	4,220	51 05	115 07	
Station............................	4,220			C. P. R.
Kananaskis river, 2·3 miles E., water 4,160; rail ..	4,202			C. P. R.
Bow river, 1·1 mile E., water 4,188; rail....	4,201			C. P. R.
Kananaskis lake, Lower...............	5,390	50 34	115 08	Wilcox
Kananaskis lake, Upper...............	5,454	50 36	115 10	Wilcox
Kananaskis pass......................	7,862	50 39	115 18	Wilcox
Kananaskis range....................	8,000-9,000	50 45	115 15	Geol. Surv.
Keith................................		51 06	114 13	
Station............................	3,553			C. P. R.
Bow river, 2·7 miles E., water 3,092; rail	3,506			C. P. R.
Kininvie.............................	2,420	50 24	111 30	C. P. R.
Kipp.................................	3,238	49 41	113 04	C. P. R.
Knee hills...........................	3,100	51 40	113 28	Geol. Surv.
Kootenay pass, South, summit (watershed range) ...	7,100	49 07	114 08	Geol. Surv.
Kootenay pass, North do	6,750	49 23	114 36	Geol. Surv.
Lacombe.............................	2,783	52 27	113 45	C. P. R.
Laggan..............................	5,037	51 26	116 12	C. P. R.
Langdon.............................	3,291	50 58	113 43	C. P. R.
Lathom..............................	2,559	50 43	112 20	C. P. R.
Leavings............................		49 51	113 31	
Station............................	3,263			C. P. R.
Willow creek, 6·1 miles S., water 3,153; rail.....	3,153			C. P. R.
Leduc...............................	2,381	53 17	113 32	C. P. R.
Lefroy, Mount.......................	11,080	51 21	116 16	Interior
Lethbridge..........................		49 42	112 50	
Station............................	2,982			C. P. R.
Summit, 3·4 miles E.................	3,014			C. P. R.
Montana junction...................	3,000			C. P. R.
Crowsnest junction.................	2,997			C. P. R.
B.M., N.E. corner of court house....	2,982·8			Irrig. Surv.
Little Fish lake....................	2,915	51 23	112 13	Geol. Surv.
Little Fork pass....................	6,775	51 41	116 27	C. P. R.
Livingstone (now Burmis)............	3,095	49 35	114 16	C. P. R.
Livingstone range..................	7,000-8,000	50 14	114 30	Geol. Surv.
Louise, Lake.......................	5,675	51 25	116 13	Interior
Lyell, Mount.......................	12,000	51 56	117 05	Collie
McDougall, Mount...................	8,500	50 56	115 04	Geol. Surv.
Macleod............................		49 43	113 26	
Station............................	3,128			C. P. R.
Calgary and Edmonton Junction......	3,139			C. P. R.
West Macleod station...............	3,108			C. P. R.
Oldman river, low water, 3,088; rail.	3,108			C. P. R.
Macoun, Mount.....................	8,697	51 34	116 03	Geol. Surv.
Magrath............................	3,210	49 25	112 52	St. M. R. R.
Medicine Lodge hills...............	3,525	52 28	114 15	Geol. Surv.
Mesa hill..........................	5,519	50 47	114 33	Irrig. Surv.
Middle coulee, on E. bdy., Tp. 4, R. XVIII	3,200	49 21	112 20	Irrig. Surv.
Middle Fork tank..................	3,815	49 35	114 15	C. P. R.
Midnapore..........................	3,423	50 54	114 05	C. P. R.
Miette, Roche......................	8,500	53 10	117 54	Geol. Surv.
Milk River ridge...................	4,200	49 15	112 30	Geol. Surv.
Milk River station................	3,429	49 08	112 04	A. R. & C. Co.
Miller.............................	2,469	53 07	113 26	C. P. R.
Minne-wanka or Devils lake.........	4,800	51 15	115 25	Interior
Mirror lake........................	6,530	51 25	116 14	Wilcox
Mist mountain......................	10,080	50 34	114 53	Geol. Surv.
Mitford (closed)...................	3,720	51 12	114 29	C. P. R.

ALBERTA

Locality	Elev.	Lat.		Long.		Authority
		'	"	'	"	
Montana junction	3,000	49	42	112	47	C. P. R.
Moose mountain............................	7,980	50	56	114	50	Irrig. Surv.
Morley...	4,067	51	10	114	52	C. P. R.
Morningside...............................	2,798	52	35	113	39	C. P. R.
Murchison, Mount	13,500	51	48	116	37	Collie
Mylle tank.	3,493	49	33	113	48	C. P. R.
Namaka..................................	5,042	50	56	113	14	C. P. R.
Nanton	3,350	50	21	113	47	C. P. R.
Nigger John hill..	5,187	50	42	114	27	Irrig. Surv.
Niles, Mount..............................	9,510	51	30	116	25	Interior
North Fork pass...........................	6,773	49	55	114	43	Geol. Surv.
Nose hill.................................	3,925	51	07	114	10	Geol. Surv.
Okotoks...................................		50	43	113	59	
Station.................................	3,430					C. P. R.
Sheep river, 0·9 mile E., water, 3,404; rail	3,422					C. P. R.
Okotoks hill..............................	5,638	50	37	114	23	Irrig. Surv.
Old Forget-me-not peak....................	7,624	50	46	114	48	Irrig. Surv.
Olds..	3,402	51	48	114	08	C. P. R.
Otoskwan.................................	2,367	53	25	113	30	C. P. R.
Palliser range............................	9,930	51	28	115	37	Interior
Panther mountain.........................	9,595	51	31	115	40	Interior
Peace hills...............................	2,625	53	00	113	25	Geol. Surv.
Pearce....	3,000	49	41	113	14	
Station.................................	3,000					C. P. R.
Belly river, 2·3 miles E., 2,965; rail....	2,987					C. P. R.
Peechee, Mount...........................	9,615	51	13	115	23	Interior
Peigan	3,310	49	37	113	35	C. P. R.
Penhold	2,945	52	08	113	53	C. P. R.
Pigeon lake	2,824	53	00	114	00	Geol. Surv.
Pigeon mountain..........................	7,845	51	02	115	12	Interior
Pilot mountain............................	9,680	51	11	115	49	Interior
Pincher.......................·..........	3,818	49	32	113	58	C. P. R.
Pinto lake................................	5,850	52	10	116	50	Coleman
Pipestone pass	8,300	51	46	116	16	Collie
Ponoka...................................		52	43	113	35	
Station.................................	2,633					C. P. R.
Battle river, water, 2,612; rail..........	2,627					C. P. R.
Pope peak..	9,625	51	25	116	17	Interior
Porcupine hills...........................	5,800	50	00	114	00	Geol. Surv.
Prairie Creek hill....	7,224	50	53	114	49	Irrig. Surv.
Prow mountain...........................	8,683	51	08	115	54	Geol. Surv.
Purple Springs............................	2,644	49	49	111	58	C. P. R.
Pyramid mountain.........................	10,800	51	51	116	42	Collie
Pyramid mountain.	9,000	52	57	118	09	Geol. Surv.
Quill lakes................................	2,860	52	05	113	10	Geol. Surv.
Radnor...................................	3,885	51	13	114	40	C. P. R.
Rae, Mount	10,160	50	40	115	00	Geol. Surv.
Red Deer.................................		52	16	113	50	
Station.................................	2,806					C. P. R.
Red Deer river, high water, 2,784; low water, 2,773; rail.	2,798					C. P. R.
Robinson hill.............................	4,809	50	54	114	31	Irrig. Surv.
Rocky buttes..............................	3,100	50	32	112	40	Geol. Surv.
Rundle, Mount............................	9,665	51	09	115	30	Interior
St. Piran, Mount..........................	8,610	51	25	116	15	Interior
Saddle mountain..........................	7,900	51	23	116	12	Interior
Saddle peak..............................	9,278	51	14	115	14	Interior
Sarbach, Mount...........................	11,100	51	52	116	47	Collie
Sarcee butte	4,476	51	02	114	32	Irrig. Surv.
Saskatchewan, Mount	12,000	52	03	117	0	Collie
Sawback range............................	10,000	51	22	115	49	Interior
Sheep mountain...........................	7,520	49	02	113	51	Geol. Surv.
Sheol, Mount.............................	10,680	51	23	116	13	Interior
Shepard..................................	3,369	50	57	113	56	C. P. R.
Simpson pass.............................	6,650	51	02	115	41	Geol. Surv.
Snake Valley marsh........................	2,872	50	30	112	53	Geol. Surv.
Southesk...	2,501	50	39	112	10	C. P. R.
Southesk cairn...........................	8,050	52	44	117	00	Geol. Surv.

Locality	Elev.	Lat.	Long.	Authority
South Twin peak	8,444	50 49	114 57	Irrig. Surv.
Spring Coulee station	3,578	49 20	113 03	St. M. R. R.
Stavely		50 10	113 30	
Station	3,411			C. P. R.
Summit, 1.6 mile N.	3,441			C. P. R.
Stirling		49 .	112 31	
Station	3,045			A. R. & C. Co.
Summit, 8.4 miles S.	3,191			A. R. & C. Co.
Stoney Squaw mountain	6,100	51 12	115 34	Interior
Storm mountain	10,360	51 12	116 00	Interior
Strathmore	3,032	50 59	113 20	C. P. R.
Stutfield peak	11,400	52 12	117 26	Collie
Suds lake	3,150	49 26	112 24	Irrig. Surv.
Suette, Roche	8,100	53 08	118 07	Geol. Surv.
Sullivan, Lake	2,605	52 00	112 00	C. P. R.
Sullivan peak	7,838	51 56	116 58	Collie
Sulphur mountain	8,030	51 07	115 32	Interior
Summit lake	5,317	51 27	116 18	C. P. R.
Surcee butte	3,030	51 40	112 59	Irrig. Surv.
Survey peak	8,650	51 55	116 50	Collie
Temple, Mount	11,637	51 21	116 12	Interior
'The Gap' siding	4,232	5. 03	115 15	C. P. R.
'The Twins' peak	11,800	52 08	117 28	Collie
Thompson, Mount	10,700	51 39	116 29	Collie
Thompson pass	6,800	52 00	117 20	Collie
Three Point mountain	7,959	50 43	114 50	Irrig. Surv.
Three Sisters, east peak	8,840	51 01	115 20	Interior
Three Sisters, west peak	9,734	51 01	115 21	Interior
Tilley	2,462	50 27	111 39	C. P. R.
Topknot hill	5,624	50 53	114 38	Irrig. Surv.
Tunnel mountain	5,540	51 11	115 33	Interior
Turtle mountain	7,000	49 34	114 25	Geol. Surv.
Tyrrell lake	3,140	49 23	112 18	Irrig. Surv.
Tyrrell, Mount	8,846	51 42	115 51	Geol. Surv.
Verdigris lake	3,127	49 15	112 00	Irrig. Surv.
Vermilion pass	5,264	51 13	116 03	Geol. Surv.
Vermilion range	9,855	51 28	115 47	Interior
Victoria, Mount	11,150	51 22	116 18	Interior
Ware Head mountain	6,954	50 40	114 44	Irrig. Surv.
Wavy lake	2,230	52 50	112 05	Geol. Surv.
West Macleod	3,108	49 43	113 28	C. P. R.
Wetaskiwin	2,480	52 59	113 20	C. P. R.
Wetmore	2,654	49 48	112 03	C. P. R.
Whaleback mountain	6,275	50 34	114 31	Irrig. Surv.
White or Wabamun lake	2,369	52 30	114 30	C. P. R.
White, Mount	9,133	51 09	115 51	Geol. Surv.
Waoopup		49 34	112 50	
Station	2,974			C. P. R.
St. Mary river, water, 2,739; rail	2,795			C. P. R.
Whyte, Mount	10,365	51 24	116 17	Interior
Wilcox pass	7,000	52 09	117 10	Collie
Wilcox peak	10,000	52 12	117 14	Collie
Wilson, Mount	10,500	51 58	116 45	Collie
Wind mountain	10,100	50 59	115 14	Geol. Surv.
Wintering hills	3,250	51 15	112 30	Geol. Surv.
Woodpecker	2,734	49 45	112 15	C. P. R.
Woolley peak	11,700	52 17	117 32	Collie

ASSINIBOIA

Locality	Elev.	Lat.	Long.	Authority
Aikins	2,401	50 19	107 42	C. P. R.
Alameda	1,864	49 16	102 17	C. P. R.
Antelope	2,556	50 08	108 26	C. P. R.
Antelope lake	2,328	50 17	108 23	C. P. R.
Antler	1,791	49 33	101 24	C. P. R.
Arcola	1,902	49 34	102 30	C. P. R.
Aylesbury	861	50 56	105 41	C. P. R.
Balgonie	2,187	50 29	104 16	C. P. R.
Belle Plaine	1,892	50 24	105 09	C. P. R.
Bench Marks*—				Irrig. Surv.
N.E. c. of S.E. ¼ Sec. 19, Tp. 17, R. XIII, W. of 2nd Mer.	2,147·4	50 26	103 46	
N.E. cor., S.E. ¼ Sec. 32, Tp. 17, R. XIII, W. of 2nd Mer.	2,065·3	50 28	103 44	
N.E. cor., N.E. ¼ Sec. 14, Tp. 18, R. XIII, W. of 2nd Mer.	1,925·0	50 31	103 40	
Top of iron bar, N.E. cor., Tp. 16, R. VIII, W. of 2nd Mer.	1,982·2	50 23	102 58	
Rail, Can. Pac. Ry. over E. end of trestle 233, over Summerberry brook, in S.E. ¼ Sec. 12, Tp. 17, R. VIII, W. of 2nd Mer.	1,932·1	50 03	102 58	
N.E. cor., Tp. 17, R. XI, W. of 3rd Mer.	2,320·5	50 . 3	107 23	
E. bdy. Sec. 25, Tp. 18, R. XI, at Swift Current and ' Elbow ' trail crossing.	2,470·4	50 33	107 23	
S.E. cor., Tp. 19, R. XI, W. of 3rd Mer.	2,410·6	50 34	107 24	
N.E. do do do	2,300·1	50 39	107 24	
N.E. cor., Tp. 19, R. IX, W. of 3rd Mer.	2,508·5	50 39	107 07	
35 chs. W. of N.E. cor., Sec. 2, Tp. 10 R. XXIV, W. of 3rd Mer.	3,597·0	49 48	109 09	
5½ chs. N. of N.E. cor., Tp. 8, R. XXIII, W. of 3rd Mer.	3,791·9	49 42	108 59	
N.E. cor., Tp. 7, R. XXIII, W. of 3rd Mer.	3,790·6	49 37	108 59	
In N.W. ¼ Sec. 21, Tp. 6, R. XXIV, W. of 3rd Mer., on bank of Frenchman brook.	3,142·8	49 30	109 19	
45 92 chs. W. of N.E. cor., Tp. 7, R. XXV, W. of 3rd Mer.	3,900·6	49 37	109 16	
11 chs. E. of S.E. cor., Sec. 4, Tp. 7, R. XXV, W. of 3rd Mer.	3,556·8	49 31	109 19	
N.E. cor. Tp. 9, R. XXV W. of 3rd Mer.	3,543·0	49 47	109 15	
Do 9, R. XXVI do	3,178·8	49 47	109 23	
Do 11, R. XXVII do	2,537·0	49 57	109 33	
Do 9, R. XXVII do	2,928·2	49 47	109 31	
Do 4, R. XXVI do	3 78·9	49 21	109 22	
N.E. cor., Sec. 34, Tp. 4, R. XXVII, W. of 3rd Mer.	3, 08·1	49 21	109 32	
22 chs. W. of N.E. cor., Sec. 34, Tp. 7, R. XXVII, W. of 3rd Mer.	4,150·5	49 37	109 34	
N.E. cor. Tp. 11, R. XXVIII, W. of 3rd Mer.	2,547·7	49 57	109 41	
Do 9, R. XXIX do	2,974·0	49 47	109 47	
Do 11, R. XXIX do	2,630·6	49 57	109 49	
Do 12, R. XXIX do	2,434·9	50 03	109 49	
Do 9, R. XXVIII do	2,902·4	49 47	109 49	
N.E. cor., Sec. 34, Tp. 7, R. XXIX, W. of 3rd Mer.	4,495·5	49 37	109 50	
45 chs. E. of N.E. cor., Sec. 34, Tp. 4, R. XXX, W. of 3rd Mer.	3,373·6	49 21	109 56	
N.E. cor. Tp. 11, R. I, W. of 4th Mer.	2,443·8	49 57	110 00	
Do 9, R. I do	3,112·4	49 47	110 00	
Do 10, R. I do	2,664·5	49 52	110 00	
45 chs. E. of N.E. cor., Sec. 34, Tp. 4, R. II, W. of 4th Mer.	5,576·9	49 21	110 10	
22·03 chs. S. of N.E. cor., Sec. 13, Tp. 5, R. III, W. of 4th Mer.	3,736·3	49 24	110 17	

* When not otherwise stated these benches are iron bars driven to within about 10 inches of the surface

ASSINIBOIA

Locality	Elev.	Lat.		Long.		Authority
		''	'	'	'	
Bench Marks—						Irrig. Surv.
N. E. cor., Tp. 7, R. 11, W. of 4th Mer....	4,024·6	49	37	110	06	
N. 20 cha. S. of N. E. cor., Sec. 13, Tp. 8, R. III, W. of 4th Mer....	4,577·3	49	40	110	17	
N. E. cor., Tp. 11, R. III, W. of 4th Mer.... ...	2,466·8	49	57	110	16	
Do 8, R. III do	4,005·0	49	42	110	16	
Do 9, R. III do	3,276·2	49	47	110	16	
Do 11, R. IV do	2,578·7	49	57	110	25	
Do 11, R. V do	2,488·0	49	57	110	33	
Do 11, R. VI do	2,380·0	49	57	110	41	
Bethune..............	1,819	50	42	105	24	C. P. R.
Bienfait............	1,867	49	09	102	47	C. P. R.
Big Stick lake........	2,274	50	16	109	20	Geol. Surv.
Bladworth..........		51	22	106	08	
Station............	2,067					C. P. R.
Summit, 4·2 miles N...	2,004					C. P. R.
Boharm..........	1,702	50	24	105	40	C. P. R.
Bonnington.........	2,009	51	30	106	16	C. P. R.
Bow Island station....	2,606	49	52	111	20	C. P. R.
Bowell.......		50	09	110	58	
Station...........	2,575					C. P. R.
Summit, 0·6 mile W....	2,594					C. P. R.
Bredenbury.......	1,711	50	56	102	02	C. P. R.
Broadview.......		50	22	102	33	
Station.........	1,960					C. P. R.
Weed brook, 3·8 miles W., bed 1,920; rail...	1,932					C. P. R.
Summit, 5 miles E....	2,034					C. P. R.
Buffalo Pound lake....	1,665	51	40	105	30	Hind
Bull's Head....	2,416	49	55	110 * 46		C. P. R.
Burrows....	1,948	50	18	102	07	C. P. R.
Carievale........	1,664	49	17	101	36	C. P. R.
Carlyle...	2,064	49	38	102	16	C. P. R.
Carmichael (formerly Cypress)...	2,637	50	02	108	40	C. P. R.
Carnduff.....	1,723	49	17	101	48	C. P. R.
Caron.....	1,841	50	27	105	51	C. P. R.
Chamberlain....	1,873	50	56	105	31	C. P. R.
Chaplin......	2,292	50	27	106	39	C. P. R.
Chaplin lake...	2,180	50	20	106	30	C. P. R.
Churchbridge.........	1,723	50	55	101	53	C. P. R.
Colley.....		49	57	109	16	
Station......	2,509					C. P. R.
Summit, 2·0 miles E...	2,568					C. P. R.
Condie (formerly Wascana)...	1,865	50	33	104	45	C. P. R.
Craik.......	1,906	51	02	105	48	C. P. R.
Crane Lake station ...	2,518	50	01	109	03	C. P. R.
Crane lake...	2,444	50	05	109	05	C. P. R.
Craven junction....	1,630	50	39	104	49	C. P. R.
Craven.....		50	42	104	49	
Station	1,606					C. P. R.
On Appelle river, 1·2 mile S., low water, 1,504; rail..	1,609					C. P. R.
Last Mountain lake...	1,597					C. P. R.
Crescent lake, approx..	1,658	51	06	102	25	C. P. R.
Crooked lake...	1,389	50	36	102	45	Hind
Cummings...	2,378	50	01	109	55	C. P. R.
Cypress hills —						
Summit.	4,243	49	38	109	35	Geol. Surv.
"The Gap"...	3,745	49	36	109	41	Geol. Surv.
Summit...	4,790	49	37	110	15	Geol. Surv.
"Head of Mountain"...	4,273	49	37	110	23	Geol. Surv.
Cypress lake.......	3,264	49	28	109	30	Geol. Surv.
Devil lake....	1,911	51	03	107	07	Geol. Surv.
Disley......		50	38	105	12	
Station........ ...	1,819					C. P. R.
Summit, 0·9 mile S...	1,853					C. P. R.
Drinkwater....	1,896	50	18	105	08	C. P. R.
Dry lake....	2,031	49	19	109	10	Geol. Surv.
Dundurn......	1,724	51	49	106	30	C. P. R.
Dunmore junction.	2,308	49	59	110	37	C. P. R.

ASSINIBOIA

Locality	Elev.	Lat.		Long.		Authority
Dunmore station	2,405	49	58	110	35	C. P. R.
Elkwater lake	4,044	49	40	110	18	Geol. Surv.
Ernfold	2,288	50	28	106	51	C. P. R.
Estevan	1,860	49	08	102	59	C. P. R.
Eyebrow or Sandhill lake	1,755	51	56	106	10	Hind
Findlater	1,810	50	51	105	24	C. P. R.
Finsbury		51	14	105	58	
Station	1,970					C. P. R.
Summit, 1·8 mile E.	2,039					C. P. R.
Fishing lakes	1,500	50	45	103	45	Hind
Fleming	1,794	50	04	101	30	C. P. R.
Forres	2,428	50	03	109	49	C. P. R.
Fort Walsh	3,624	49	35	109	53	Geol. Surv.
Frobisher	1,883	49	13	102	25	C. P. R.
Gainsborough		49	17	101	25	
Station	1,602					C. P. R.
North Antler brook, water 1.594 ; rail	1,604					C. P. R.
Girvin	1,965	51	09	105	55	C. P. R.
Glen Ewen	1,817	49	13	102	00	C. P. R.
Goose Lake station (closed)	2,405	50	12	108	10	C. P. R.
Grand Coulée station	1,857	50	26	104	48	C. P. R.
Green lake	3,557	49	28	110	01	Geol. Surv.
Grenfell	1,957	50	24	102	55	C. P. R.
Gull Lake station	2,562	50	06	108	29	C. P. R.
Halbrite	1,894	49	28	103	32	C. P. R.
Hanley	1,869	51	39	106	29	C. P. R.
Hay lake	3,754	49	39	109	25	Geol. Surv.
Herbert	2,311	50	25	107	14	C. P. R.
Highpound or Buffalo Pound lake	1,665	51	40	105	30	Hind
Hirsch station	1,874	49	11	102	36	C. P. R.
Indian Head	1,924	50	32	103	40	C. P. R.
ine		49	57	110	16	
Station	2,493					C. P. R.
Summit, 1·9 mile E.	2,522					C. P. R.
Johnston lake	2,189	50	00	106	00	C. P. R.
Kincorth	2,531	49	58	109	40	C. P. R.
Langenburg	1,660	50	50	101	41	C. P. R.
Langevin	2,495	50	17	111	21	C. P. R.
Last Mountain or Long lake	1,597	51	00	105	15	C. P. R.
Leech lake	1,658	51	05	102	28	C. P. R.
Leven	2,468	50	15	108	00	C. P. R.
Lumsden		50	38	104	52	
Station	1,620					C. P. R.
Qu'Appelle river, 3·5 miles W., water 1,606 ; rail	1,625					C. P. R.
McLean	2,284	50	30	104	04	C. P. R.
Macoun	1,891	49	19	103	16	C. P. R.
Manor	2,066	49	36	102	05	C. P. R.
Many Island lake	2,304	50	07	110	05	Geol. Surv.
Maple Creek		49	55	109	28	
Station	2,495					C. P. R.
B.M., corner stone of Dixon Bros. store, N.W. corner	2,496·8					Irrig. Surv.
Maple creek, high water, 2,491 ; rail	2,495					C. P. R.
Medicine Hat		50	02	110	40	
Station	2,171					C. P. R.
South Saskatchewan river, low water, 2,137 ; high water, 2,154 ; rail	2,173					C. P. R.
Milestone	1,903	50	00	104	30	C. P. R.
Missouri coteau—						
Near south end of Chaplin lake	2,395	50	20	106	10	Geol. Surv.
At Secretan	2,283	50	28	106	25	Geol. Surv.
Vermilion hills	2,255	50	42	106	55	Geol. Surv.
North of S. Saskatchewan river	2,347	50	47	107	26	Geol. Surv.
International boundary	3,200	49	00	106	11	Geol. Surv.
Moosejaw		50	22	105	32	
Station	1,767					C. P. R.
Moosejaw brook, bed, 1.740 ; rail	1,761					. P. R.
Moosomin	1,884	50	08	101	40	C. P. R.
Morse	2,274	50	24	107	02	C. P. R.

ASSINIBOIA

Locality	Elev.	Lat.		Long.		Authority.
Mortlach	1,961	50	28	106	03	C. P. R.
North Portal	1,944	49	00	102	33	C. P. R.
Oakshela	1,952	50	23	102	42	C. P. R.
Old-man-on-his-back plateau	3,465	49	12	109	18	Geol. Surv.
Old Wives lakes	2,189	50	20	106	30	C. P. R.
Outer Rainy hill	2,700	50	27	111	18	Geol. Surv.
Oxbow		49	14	102	10	
Station	1,895					C. P. R.
Summit, 0·3 mile W	1,900					C. P. R.
Pakowki lake	2,735	49	20	111	00	Geol. Surv.
Parkbeg	1,982	50	28	106	15	C. P. R.
Pasqua	1,872	50	22	105	23	C. P. R.
Pense	1,881	50	24	104	59	C. P. R.
Perceval	2,038	50	21	102	24	C. P. R.
Pilot Butte station	2,016	50	28	104	25	C. P. R.
Pinto Horse butte	3,331	49	23	107	29	Geol. Surv.
Pipestone lake	2,071	50	20	102	56	Irrig. Surv.
Qu'Appelle	2,134	50	32	103	52	C. P. R.
Red Jacket	1,917	50	12	101	47	C. P. R.
Redvers	1,935	49	34	101	43	C. P. R.
Regina		50	27	104	36	
Station	1,885					C. P. R.
Waskana river, rail	1,861					C. P. R.
Roche Percee	1,736	49	04	102	46	C. P. R.
Rokeby	1,664	51	08	102	20	C. P. R.
Rouleau	1,878	50	11	104	55	C. P. R.
Round lake	1,364	50	32	102	20	Hind
Rush Lake station	2,301	50	25	107	25	C. P. R.
Rush lake	2,293	50	25	107	21	Geol. Surv.
Saltcoats	1,714	51	02	102	10	C. P. R.
Sandhill or Eyebrow lake	1,755	51	56	106	10	Hind
Secretan		50	28	106	27	
Station	2,276					C. P. R.
Summit, 0·3 mile E	2,282					C. P. R.
Seven Persons	2,482	49	52	110	55	C. P. R.
Seward	2,474	50	13	108	05	C. P. R.
Sidewood	2,478	50	04	108	50	C. P. R.
Sintaluta	1,984	50	28	103	27	C. P. R.
Stair	2,431	50	05	110	49	C. P. R.
Suffield		50	13	111	11	
Station	2,455					C. P. R.
Depression, 4·0 miles E	2,384					C. P. R.
Summerberry	1,938	50	25	103	05	C. P. R.
Swift Current	2,423	50	16	107	48	C. P. R.
Tompkins	2,614	50	04	108	47	C. P. R.
Twelve-mile lake	2,479	49	28	106	15	Geol. Surv.
Vermilion hills	2,255	50	42	106	55	Geol. Surv.
Waldeck	2,357	50	22	107	30	C. P. R.
Walsh	2,430	49	57	110	02	C. P. R.
Wapella	1,930	50	15	101	57	C. P. R.
Waskana (now Condie)	1,865	50	33	104	45	C. P. R.
Wauchope	2,035	49	35	101	58	C. P. R.
Webb	2,583	50	10	108	14	C. P. R.
Weyburn	1,847	49	39	103	51	C. P. R.
Whitemud plateau	3,428	49	29	108	20	Geol. Surv.
Whitewood	1,966	50	20	102	16	C. P. R.
Whitla	2,746	49	52	111	05	C. P. R.
Wild Horse or de Marron lake	2,852	49	01	110	10	Geol. Surv.
Winnifred	2,741	49	54	111	12	C. P. R.
Wolseley	1,950	50	25	103	16	C. P. R.
Wood mountain—						
East summit	3,347	49	16	106	25	Geol. Surv.
West summit	3,371	49	13	107	04	Geol. Surv.
Wood Mountain, N.W.M.P. post	2,875	49	19	106	22	Geol. Surv.
Yellow Grass	1,889	49	49	104	10	C. P. R.
Yorkton	1,633	51	12	102	27	C. P. R.

ATHABASKA

Locality	Elev.	Lat.		Long.		Authority
		°	′	°	′	
Assiniboine, old Fort, water in Athabaska river.........	2,000	54	19	115	23	Geol. Surv.
Athabaska, Lake	690	59	00	110	00	Geol. Surv.
Athabaska Landing............		56	18	112	40	
Water in Athabaska river....	1,650					Geol. Surv.
Plateau..........	2,000					Geol. Surv.
Birch mountain, north.....	2,100	58	00	112	00	Geol. Surv.
Birch mountain, south	2,300	57	15	112	30	Geol. Surv.
Black Bear Island lake.	1,200	55	40	105	45	Geol. Surv.
Black lake....... ..	1,000	59	15	105	30	Geol. Surv.
Brulé lake, high water.......	3,201	53	18	117	50	C. P. R.
Buffalo Head hills..	2,500	57	30	116	00	Geol. Surv.
Buffalo lake...............	2,000	57	35	112	55	Geol. Surv.
Buffalo lake.............	1,330	56	00	109	00	Geol. Surv.
Burntwood lake..	886	55	25	100	20	Geol. Surv.
Claire lake....	700	58	40	112	00	Geol. Surv.
Clear hills.............	2,600-3,500	56	40	119	00	Geol. Surv.
Clear lake................	1,330	56	00	108	20	Geol. Surv.
Cold or Kississing lake............	900	55	15	101	15	Geol. Surv.
Cree lake.........	1,530	57	30	107	00	Geol. Surv.
Dunvegan, water in Peace river.	1,305	55	56	118	37	Geol. Surv.
Foster lakes........	1,600	56	35	105	30	Geol. Surv.
Frog portage .... .'	1,150	55	25	103	30	Geol. Surv.
Isle à la Crosse lake........	1,330	55	45	107	50	Geol. Surv.
Iroquois lake.... ..	1,830	55	27	117	05	Geol. Surv.
Jasper lake........	3,256	53	05	118	10	C. P. R.
Kenesis or Sturgeon lake.	2,000	55	07	117	50	Geol. Surv.
Kidney lake........	2,320	56	46	114	35	Geol. Surv.
Kississing or Cold lake............	900	55	15	101	15	Geol. Surv.
Knee lake....	1,250	55	50	107	10	Geol. Surv.
La Loche lake............	1,577	56	30	109	35	Geol. Surv.
Lesser Slave lake.	1,890	55	20	115	30	Geol. Surv.
Lesser Slave plateau.	3,090	55	30	114	45	Geol. Surv.
Long lake........	2,269	56	44	114	42	Geol. Surv.
Loon lake.	1,660	56	33	115	30	Geol. Surv.
Martin mountain............	2,900	55	30	114	45	Geol. Surv.
McMurray, Fort, water in Athabaska river......... ..	840	56	42	111	30	Geol. Surv.
Methy portage, summit.......	1,705	56	40	109	55	Geol. Surv.
Do 'Cockscomb'.	1,743	56	40	109	55	Geol. Surv.
Peace River Landing, water in Peace river...........	1,225	56	15	117	15	Geol. Surv.
Pelican lake............	1,910	55	48	113	25	Geol. Surv.
Reindeer lake..... .:	1,150	57	30	102	30	Geol. Surv.
Sandy lake............	1,910	55	50	113	30	Geol. Surv.
Sturgeon or Kenesis lake....	2,000	55	07	117	50	Geol. Surv.
Trout lake.	2,130	56	35	114	40	Geol. Surv.
Trout mountain...... .... •	2,350	56	45	114	30	Geol. Surv.
Vermilion, Fort, water in Peace river.	950	58	24	115	58	Geol. Surv.
Wabiskaw lake, Lower	1,705	55	57	113	55	Geol. Surv.
Wabiskaw lake, Upper........	1,720	56	05	114	05	Geol. Surv.
Whitefish lake....	2,075	55	50	115	30	Geol. Surv.
Wollaston lake.....	1,300	58	15	103	30	Geol. Surv.

BRITISH COLUMBIA

Locality	Elev.	Lat.		Long.		Authority
		°	′	°	′	
Aaron hill....	5,100	49	17	117	32	Geol. Surv.
Abbotsford....	89	49	02	122	14	C.P.R.
Abbott ridge....	7,710	51	20	117	29	Interior
Aberdeen mountain..........	6,100	50	21	119	04	Geol. Surv.
Adair, Mount.........	2,889	49	35	126	31	Admir. chart
Adams lake	1,357	51	10	119	35	Geol. Surv.
Adams, Mount	1,610	50	57	127	10	Admir. chart
Adams plateau	6,000	51	03	119	34	Geol. Surv.
Addenbroke, Mount....	5,140	50	14	124	42	Admir. chart
Afton mountain.........	8,423	51	14	117	31	Interior
Agassiz		49	14	121	46	C.P.R.
Station......................	60					C.P.R.
Fraser river, 3·1 miles W., high water, 45; ordinary water	30					C.P.R.
Fraser river, 1·9 mile E., high water, 61; ordinary water	54					C.P.R.
Agnes, Mount..........	6,200	52	58	121	36	Geol. Surv.
Aiken, Mount.........	5,940	50	17	124	35	Admir. chart
Airy mountain	8,540	49	33	117	50	Geol. Surv.
Akamina boundary monument..	7,524	49	00	114	03	Bdy. Com., Can.
Akasik mountain..........................	8,660	50	14	121	45	Geol. Surv.
Akayihi, Mount.......................	7,100	56	16	123	16	Carruthers
Alamo Concentrator siding.............	2,452	50	00	117	18	C.P.R.
Albemarle, Mount	3,756	49	35	126	28	Admir. chart
Albert Cañon station	2,227	51	09	117	52	C.P.R.
Albert, Mount........................	8,260	50	15	123	52	Admir. chart
Albert Edward, Mount................	6,968	49	41	125	27	Admir. chart
Albreda lake, high water.............	2,861	53	37	119	08	C.P.R.
Aldridge (formerly Moyelle).......	3,051	49	16	115	50	C.P.R.
Alexandra.........		49	06	123	54	
Station	128					E. & N.R.
Nanaimo river, 2·0 miles S., freshet, 68; low water, 43; rail.	133					E. & N.R.
Alexandra peak	6,394	49	44	125	30	Admir. chart
Alfred, Mount........................	8,450	50	14	124	06	Admir. chart
Alice, Mount (Jervis inlet)....	6,285	50	12	124	04	Admir. chart
Alice, Mount (Laredo channel).......	1,150	52	46	128	58	Admir. chart
Allison or South pass	5,808	49	15	120	47	Geol. Surv.
Anchor, Mount.......................	1,981	53	52	130	15	Admir. chart
Anderson lake, high water, 858; low water	853	50	35	122	25	C.P.R.
Anderson, Mount.....................	5,400	53	00	121	50	Geol. Surv.
Annesley, Mount..	2,760	51	04	127	13	Admir. chart
Anthony, Mount......................	1,100	50	33	126	15	Admir. chart
Antler mountain.....................	5,700	52	57	121	30	Geol. Surv.
Antony, Mount	5,069	50	42	126	07	Admir. chart
Anvil mountain	7,500	59	00	129	55	Geol. Surv.
Anvil peak	3,279	49	58	127	09	Admir. chart
Arbutus summit	1,702	49	13	124	51	Admir. chart
Arctic Pacific divide, Summit lake to Salmon river.... ..	2,500	54	14	122	45	Geol. Surv.
Arctic-Pacific divide, McLeod and St. James trail.... ..	2,820	54	45	123	30	Geol. Surv.
Arctic-Pacific divide.................	2,730	58	26	130	02	Geol. Surv.
Armstrong		50	27	119	11	C.P.R.
Station	1,187					C.P.R.
Summit, 1·0 mile N	1,196					C.P.R.
Arrowhead............................	1,413	50	40	117	54	C.P.R.
Arrow lake, Upper, high water (1894), 1,419; low water..	1,384	50	30	117	55	C.P.R.
Arrow lake, Lower, high water, 1,416; low water.......	1,382	49	30	118	00	C.P.R.
Arrowsmith, Mount.	5,976	49	13	124	36	Admir. chart
Arrowstone hills, north.....	5,700	51	00	121	09	Geol. Surv.
Arrowstone hills, south	5,740	50	55	121	13	Geol. Surv.

BRITISH COLUMBIA

Locality	Elev.	Lat.		Long.		Authority
		°	′	°	′	
Arthur, Mount (Pender harbour)	1,500	49	38	124	02	Admir. chart
Arthur, Mount (Jervis inlet)	5,585	50	09	123	56	Admir. chart
*Arthur peak	3,588	58	03	133	58	Bdy. Com., Can.
Arthurs Seat	5,500	50	25	120	25	Geol. Surv.
Ashby mountain, north	5,173	55	39	129	54	Admir. chart
Ashby mountain, south	5,485	55	36	129	54	Admir. chart
Ashcroft		50	43	121	17	
Station	1,004					C.P.R.
Thompson river at mouth of Bonaparte river, high water 966 ; low water	959					C.P.R.
Ashington range	4,041	55	08	130	01	Admir. chart
Askom mountain	8,150	50	29	121	49	Geol. Surv.
Assiniboine, Mount	11,860	50	56	115	42	Interior
Asulkan pass	7,063	51	11	117	28	
Athabaska pass	6,025	52	27	118	20	C.P.R.
Do	5,710					Coleman
Atlin lake	2,200	59	30	133	45	Geol. Surv.
Atlin mountain	5,500	59	33	133	52	Geol. Surv.
Atna mountains	9,000	55	40	127	20	Geol. Surv.
Atna pass	6,000	55	47	127	15	Geol. Surv.
Avalanche, Mount	9,365	51	17	117	27	Interior
Babine range	7,000-8,000	55	30	127	30	Geol. Surv.
Babine lake	2,222	54	45	126	00	Geol. Surv.
Badger peak	6,500	52	38	121	20	Geol. Surv.
Beothaw, Mount	2,800	50	33	125	29	Admir. chart
*Bainbridge peak	3,467	57	06	132	59	Coast Surv.
----er, Mount	7,100	49	28	115	37	Geol. Surv.
Baker pass	6,700	51	45	116	43	Collie
Bald mountain	2,925	50	51	126	28	Admir. chart
Bald peak	3,330	50	49	125	35	Admir. chart
Balfour Knob	7,406	49	40	117	00	Geol. Surv.
Balfour, Mount	10,000	51	34	116	28	Interior
Bare mountain	1,500	50	31	126	14	Admir. chart
Bareside mountain	1,645	53	55	130	18	Admir. chart
Baresides, Mount	3,010	50	32	125	58	Admir. chart
Barham peak	6,820	59	46	133	20	Geol. Surv.
Barker mountain	6,500	52	45	121	15	Geol. Surv.
Barner, Mount	6,090	50	27	124	37	Admir. chart
Barnett	22	49	17	122	56	C.P.R.
Basil Lump	2,960	54	30	130	21	Admir. chart
Basque	895	50	39	121	18	C.P.R.
Bastion mountains	5,360	50	50	119	09	Geol. Surv.
Bate, Mount	5,091	49	52	126	30	Admir. chart
Bauke, Mount	2,018	49	34	126	33	Admir. chart
Bazalgette range	5,300	50	40	125	37	Admir. chart
B. C. mine—Summit camp	3,800	49	08	118	31	C.P.R.
Bear Creek station		50	21	117	27	
Station	3,673					C.P.R.
Stony creek, 1·5 mile E., bed, 3,224 ; rail	3,501					C.P.R.
Bear Lake station		50	02	117	11	
Station	3,500					K. & S.R.
Summit	3,515					K. & S.R.
Bear lake	3,507	50	02	117	11	K. & S.R.
Bear lake	2,523	53	55	122	00	C.P.R.
Bear lake	2,900	56	00	126	45	Rys. & Canals
Bear mountain	3,500	53	59	122	08	Geol. Surv.
Beatrice peak	3,410	50	43	125	51	Admir. chart
Beaver siding	1,984	49	07	117	32	N. & F. S. R.
Beaver mountain	6,400	49	08	117	26	Geol. Surv.
Beaverfoot range	8,000-9,000	51	10	116	40	Geol. Surv.
Beavermouth		51	31	117	24	
Station	2,435					C.P.R.
Beaver river, bed, 2,529 ; rail	2,559					C.P.F
Beckford, Mount	3,500	50	58	126	33	Admir. chart
Bedingfield range	3,800	49	20	125	51	Admir. chart
Bee peak	6,545	59	31	134	09	Geol. Surv.
Beehive mountain	8,500	50	35	114	44	Geol. Surv.
Begbie, Mount	8,842	50	53	118	17	Interior

* All peaks marked with an asterisk are in the disputed Yukon-Alaska territory

BRITISH COLUMBIA

Locality	Elev.	Lat.		Long.		Authority
Ben mountain	2,460	54	32	130	22	Admir. chart
Benedict, Mount	3,500	51	02	126	28	Admir. chart
Bennett, Lake	2,161	60	00	134	55	W. P. & Y. R.
Bennett station	2,170	59	51	134	55	W. P. & Y. R.
Benson, Mount	3,366	49	09	124	04	Admir. chart
Beaufort range	5,420	49	28	125	00	Admir. chart
Berkeley, Mount	3,887	50	34	125	45	Admir. chart
Big Bar lake	3,630	51	19	121	47	Geol. Surv.
Big Bar P. O.	3,300	51	18	122	01	Geol. Surv.
*Big mountain	8,750	56	54	131	35	Admir. chart
Birch mountains	6,750	59	26	133	49	Geol. Surv.
*Black Crag	5,900	56	33	131	42	Admir. chart
Black mountain (Howe sound)	4,280	49	23	123	14	Admir. chart
*Black mountain (Lynn canal)	5,130	59	07	136	06	Coast Surv.
Black mountain (Atlin)	5,700	59	57	133	43	Geol. Surv.
Black Stuart mountain	6,000	52	52	121	11	Geol. Surv.
Blair, Mount	6,550	51	07	125	41	Admir. chart
Blake, Mount	3,000	50	42	125	32	Admir. chart
Blanchard, Mount or Golden Ears	5,560	49	22	122	31	Admir. chart
Blenheim Mount	2,408	48	55	124	57	Admir. chart
Bluebell mountain	7,135	49	45	116	47	Geol. Surv.
Blue ridge	6,020	50	04	117	07	Geol. Surv.
Bluff lake	2,865	51	45	124	46	C. P. R.
Blunt mountain	4,000	55	14	127	12	Geol. Surv.
Blustry mountain	7,640	50	36	121	40	Geol. Surv.
Bobbie Burns mountain	5,000	50	26	118	50	Geol. Surv.
Bonaparte lake	3,760	51	15	120	30	Geol. Surv.
Bonney, Mount	10,205	51	11	117	32	Interior
Bonnington Falls	1,658	49	28	117	30	C. P. R.
Bonwick, Mount	1,600	54	32	130	51	Admir. chart
Boofus mountain	5,200	50	56	133	02	Geol. Surv.
Bor, Mount	9,073	49	50	117	44	Geol. Surv.
Botanie mountain	6,620	50	22	121	36	Geol. Surv.
Bouleau mountain	6,000	50	19	119	41	Geol. Surv.
*Boundary mountain	4,805	57	04	131	51	Admir. chart
Bowman, Mount	7,500	51	15	121	53	Geol. Surv.
Boyds (70-mile House)	3,800	51	19	121	23	Geol. Surv.
Boyle, Mount	2,000	50	31	125	43	Admir. chart
Bradshaw, Mount	2,300	50	28	125	35	Admir. chart
Brew, Mount (Kamloops district)	7,300	50	37	121	55	Geol. Surv.
Brew, Mount (Cariboo district)	6,000	52	35	121	17	Geol. Surv.
Brisco range	7,500-8,000	50	45	116	15	Geol. Surv.
Brougham, Mount	2,203	50	24	125	27	Admir. chart
Broughton peaks, north	3,414	49	00	125	14	Admir. chart
Broughton peaks, south	3,070	48	59	125	14	Admir. chart
Brown Dome	6,575	59	49	132	58	Geol. Surv.
Brown, Mount (Athabaska pass)	9,050	52	27	118	22	Coleman
Brown, Mount (Portland canal)	5,800	55	43	129	57	Admir. chart
Browne, Mount	1,745	50	51	126	47	Admir. chart
Bruce, Mount	2,329	48	48	123	30	Admir. chart
Brunswick, Mount	6,265	49	30	123	12	Admir. chart
Bryce, Mount	13,000	51	58	117	20	Collie
Bullock, Mount	2,180	50	57	127	07	Admir. chart
Bunsen, Mount	4,100	50	17	124	47	Admir. chart
Bunster hills	3,50	50	00	124	21	Admir. chart
Burdett, Mount	6,000	52	56	121	33	Geol. Surv.
Burnaby lake	48	49	15	122	56	Hill
Burnaby range	2,500	50	55	126	42	Admir. chart
Burnett mountain	7,500	49	17	116	52	Geol. Surv.
Burns mountain	5,800	53	01	121	41	Geol. Surv.
Burnt hill	1,695	50	28	128	01	Admir. chart
Burnt mountain	2,850	50	34	1?9	11	Admir. chart
Burrell, Mount	1,500	50	55	126	40	Admir. chart
Bury, Mount	3,...	49	35	126	30	Admir. chart
Bush peak	13,000	51	52	117	12	Collie
Buttle, Mount	5,250	49	01	124	34	Admir. chart

2

BRITISH COLUMBIA.

Locality	Elev.	Lat.	Long.	Authority
		° ′	° ′	
Buxton, Mount....	3,430	51 36	128 00	Admir. chart
Cairn mountain........	7,650	50 38	121 40	Geol. Surv.
Calder, Mount......	4,960	49 54	124 00	Admir. chart
Calvert, Mount...	1,135	50 54	126 50	Admir. chart
Cambridge, Mount..	5,036	50 03	124 02	Admir. chart
Camelsfoot mountains..	7,300	50 52	122 00	Geol. Surv.
Camel Hump mountain..	4,180	50 14	118 50	Geol. Surv..
Cameron, Mount........	5,105	50 24	134 03	Geol. Surv.
Campania island.......	2,000	53 05	129 23	Admir. chart
Camp peak..	3,500	50 39	125 37	Admir. chart
Campbell mountain (Cariboo district).	5,200	52 58	121 51	Geol. Surv.
Campbell mountains (Portland canal).	4,406	55 31	129 43	Admir. chart
Canim lake...	2,557	51 50	120 50	C. P. R.
Canoe mountain..	5,500	55 30	122 38	Geol. Surv.
Cap Cone...	5,560	50 43	125 37	Admir. chart
Cariboo lake..	2,550	52 45	121 25	Geol. Surv.
Cariboo mountain.....	6,300	52 49	121 42	Geol. Surv.
*Carmack, Mount...	6,900	59 34	135 15	Bdy. Com., Can.
Carnarvon, Mount.	8,876	51 21	116 39	Interior
Carp lake..	2,747	54 50	123 25	Geol. Surv.
Carrier peak..	5,500	59 53	133 36	Geol Surv.
Carrington hills.	1,000	50 46	125 25	Admir. chart
Carruthers, Mount..	10,100	56 22	126 10	Carruthers
Cartier, Mount..	8,562	50 54	118 03	Interior
Cascade..		49 01	118 12	
Station..	1,587			C. P. R.
Kettle river, high water, 1,469 ; low water, 1,457 ; rail.	4,566			C. P. R.
Castle, Mount....	4,640	50 27	127 09	Admir. chart
Castlegar..	1,418	49 18	117 40	C. P. R.
Castor, Mount...	9,069	51 12	117 29	Interior
Catface mountains, north.	3,370	49 16	125 59	Admir. chart
Catface mountains, south..	2,891	49 16	126 00	Admir. chart
Cathedral mountain..	10,284	51 24	116 23	Interior
Catherine, Mount...	4,420	50 44	125 46	Admir. chart
Cedar..	3,188	51 25	117 30	C. P. R.
Chaloner range...	4,681	55 30	129 34	Admir. chart
Champion..	2,008	49 08	117 30	N. & F. S. R.
Champion lakes....	3,550	49 11	117 37	Geol. Surv.
Chancellor peak...	10,780	51 12	116 30	Interior
Chaperon lake...	3,060	50 12	120 03	Geol. Surv.
Chatfield island.	920	52 15	128 07	Admir. chart
Chaumox..	568	49 56	121 28	C. P. R.
Chemainus....	103	48 56	123 43	E. & N. R.
Cheops, Mount ..	8,506	51 17	117 33	Interior
Cherry Creek station..	1,142	50 43	120 38	C. P. R.
Chestatta lake..	2,800	53 42	125 15	Geol. Surv.
Chikoida mountain..	6,050	59 14	133 00	Geol. Surv.
*Chilkat peak...	6,380	59 30	136 08	Bdy. Com., Can.
Chilkat pass...	3,502	59 42	135 13	Geol. Surv.
Chilliwak lake..	2,052	49 05	121 25	Bdy. Com.
Chilliwak mountain..	6,570	49 05	121 36	Geol. Surv.
Chipooin mountain..	7,070	50 44	121 43	Geol. Surv.
Chizicnt lake..	3,254	52 20	124 03	C. P. R.
Choelquoit lake..	3,850	51 42	124 15	Geol. Surv.
Choowhels mountains..	6,200	50 32	120 35	Geol. Surv.
Christina lake..	1,531	49 08	118 14	Bdy. Com.
Chuchihum pass..	4,719	49 00	121 18	Bdy. Com.
Churchill, Mount..	6,570	49 59	123 52	Admir. chart
City of London mine..	4,150	49 01	118 36	C. P. R.
Cinder hill..	5,070	50 34	121 08	Geol. Surv.
Clanwilliam ..	1,801	50 58	118 26	C. P. R.
Clarence, Mount..	3,015	50 27	125 43	Admir. chart
Clashmore, Mount..	5,502	55 29	129 54	Admir. chart
Clearwater lake...	2,480	52 20	120 15	C. P. R.
Cliff mountain...	2,375	50 54	126 47	Admir. chart
Clinton, Junction creek at trail crossing..	3,040	51 05	121 34	Geol. Surv.
Coal Creek station..	3,221	49 26	115 02	C. P. R.
Coal hill..	3,470	50 37	120 22	Geol. Surv.
Coast Nipple ..	1,350	51 11	127 40	Admir. chart

BRITISH COLUMBIA

Locality	Elev.	Lat.			Long.			Authority
		°	′	″	°	′	″	
Coast peak	1,250	51	21		127	37		Admir. chart
Cobble Hill station	309	48	42		123	39		E. & N. R.
Cody junction	3,515	49	59		117	15		K. & S. R.
Collie, Mount	10,500	51	35		116	35		Outram
Colling range	4,000	55	50		130	00		Admir. chart
Collingwood, Mount	3,255	50	33		126	02		Admir. chart
Colnett, Mount	2,616	49	11		125	51		Admir. chart
Columbia, Mount, about	14,000*	52	05		117	32		Collie
Colville range	2,500	50	56		126	47		Admir. chart
Compton island	950	54	39		130	25		Admir. chart
Cone mountain	2,400	52	40		128	34		Admir. chart
Connor mountain	7,050	49	24		117	09		Geol. Surv.
Conolly, Mount	3,965	50	54		126	43		Admir. chart
Conspicuous peak	2,188	48	28		123	40		Admir. chart
Constable, Mount	5,825	50	57		126	22		Admir. chart
Comma peak	4,880	48	50		126	18		Admir. chart
Cooper, Mount	3,000	51	03		126	53		Admir. chart
Copper mountain (Kootenay lake)	8,300	49	49		116	31		Geol. Surv.
Copper mountain (Nelson)	7,460	49	24		117	07		Geol. Surv.
Coquitlam lake	445	49	25		122	47		Hill
Cornwall hills	6,690	50	41		121	26		Geol. Surv.
Cortes island	1,062	50	05		124	57		Admir. chart
Coryell	3,135	49	08		118	06		C. P. R.
Cosmos heights	5,746	50	35		125	04		Admir. chart
*Cosmos range	4,723	57	00		132	43		Bdy. Com., Can.
Costello peak	5,480	51	04		125	31		Admir. chart
Cougar mountain	7,882	51	16		117	36		Interior
Cowichan	128	48	44		123	40		E. & N. R.
Cox island	1,047	50	47		128	37		Admir. chart
Craig hills	3,000	50	44		126	21		Admir. chart
Craigellachie		50	58		118	46		C. P. R.
Station	1,229							C. P. R.
Eagle river, 3 1 miles W., bed, 1,178; rail	1,199							C. P. R.
Cranberry lake	2,580	52	48		119	15		C. P. R.
Cranbrook	3,014	49	32		115	46		C. P. R.
Crater mountain	6,000	59	15		132	10		Geol. Surv.
Craven hill	1,000	50	27		125	56		Admir. chart
Crawford, Mount	7,700	49	46		116	44		Geol. Surv.
Crawshay, Mount	5,500	50	11		124	34		Admir. chart
Creston junction	1,853	49	09		116	02		C. P. R.
Creston station		49	05		116	30		
Station	1,985							C. P. R.
Summit, 2 8 miles E	2,107							C. P. R.
Cridge, Mount	5,890	50	57		125	27		Admir. chart
*Crillon, Mount	12,750	58	40		137	10		Bdy. Com., Can.
Crownest	4,438	49	38		114	42		C. P. R.
Crownest pass								
Railway summit	4,449	49	38		114	42		C. P. R.
Trail summit, watershed range	4,845	49	33		114	44		Geol. Surv.
West summit, on trail	5,500	49	29		114	54		Geol. Surv.
Crown lake	2,669	50	50		121	41		C. P. R.
Crown, Mount (Vancouver island)	6,082	49	56		125	48		Admir. chart
Crown, Mount (Howe sound)	4,708	49	25		123	09		Admir. chart
Crown Point	2,638	49	04		117	46		C. P. R.
Cust house, water in Peace river	1,792	56	02		122	15		Geol. Surv.
Daly, Mount	10,255	51	31		116	23		Interior
*D'Agelet, Mount	9,550	58	36		137	11		Bdy. Com., Can.
Daghaodeya lake	2,700	56	30		125	05		Carruthers
*Dahlgren peak	3,569	57	14		133	22		Bdy. Com., Can.
Daisy lake	1,187	50	00		123	10		C. P. R.
*Dana peak	4,260	57	08		132	54		Coast. Surv.
David, Mount	800	48	48		123	11		Admir. chart
Davie Mountain	7,817	49	29		116	42		Geol. Surv.
Dawson, Mount	11,110	51	08		117	25		Interior
Dawson peak, north	6,450	59	59		132	30		Geol. Surv.

* Probably overestimated.

2½

BRITISH COLUMBIA

Locality	Elev.	Lat.		Long.		Authority
Dawson peak, south..	6,540	59	57	132	30	Geol. Surv.
Dease lake.	2,600	58	40	130	05	Geol. Surv.
Deer Mound.	2,230	54	26	130	23	Admir. chart
Deer Park mountain.	3,180	49	29	118	03	Geol. Surv.
Deltaform, Mount.	10,945	51	18	116	15	Interior
Demers, Mount.	8,640	49	53	117	40	Geol. Surv.
Denman island	535	49	31	124	48	Admir. chart
Denman, Mount.	6,580	50	17	124	32	Admir. chart
Dennis, Mount	7,971	50	22	116	28	Interior
Denny island	2,730	52	09	127	50	Admir. chart
Dent. Mount.	5,057	55	12	129	58	Admir. chart
Denver, Mount.	8,815	49	58	117	30	Geol. Surv.
Denver Cañon siding	2,107	50	00	117	22	C. P. R.
Deville, Mount	10,892	51	08	117	25	Interior
*Devils Thumb.	9,105	57	05	132	21	Bdy. Com., Can.
Dewdney.	30	49	09	122	12	C. P. R.
Diadem, Mount.	6,050	50	00	124	06	Admir. chart
Disella lake	3,805	59	20	131	45	Geol. Surv.
D'Israeli mountains.	7,000	56	05	129	55	Admir. chart
Dixie mountain.	5,670	59	35	133	09	Geol. Surv.
Dome Mountain (Selkirk mountains).	8,802	51	12	117	29	Interior
Dome Mountain (Rocky mountains).	11,850	52	06	117	19	Collie
Dominion mountain.	7,430	49	21	117	20	Geol. Surv.
Donald (mean of levels from lake Superior and Pacific)..	2,580	51	29	117	11	C. P. R.
Donkin, Mount....	9,529	51	07	117	30	Interior
Double hill.	1,400	50	48	126	39	Admir. chart
Douglas, Mount.	4,245	49	08	124	39	Admir. chart
Downie range, north.	4,026	50	25	125	03	Admir. chart
Downie range, south ...	4,400	50	23	125	02	Admir. chart
Dragon mountain.	5,600	53	04	121	52	Geol. Surv.
Drew, Mount (Vancouver island).	4,500	48	57	124	19	Admir. chart
Drew, Mount (Jervis inlet)	6,150	49	47	123	48	Admir. chart
Drewry, Mount.	7,817	49	20	116	52	Geol. Surv.
Drummond, Mount.	3,273	50	33	125	44	Admir. chart
Drynoch .	760	50	20	121	34	C. P. R.
Ducks.	1,156	50	39	119	57	C. P. R.
Duck Creek mountain.	5,000	52	43	121	39	Geol. Surv.
Dudley cone.	4,130	50	12	124	36	Admir. chart
Dufferin hill.	2,830	59	40	120	24	Geol. Surv.
Duncan, .		48	47	123	42	
Station..	23					E. & N.R.
Cowichan river, water, 24; rail ..	53					E. & N.R.
Duncan or Upper Kootenay lake.	1,800	50	20	117	00	Public Works
Dundas, Mount.	5,950	50	55	125	29	Admir. chart
Dunsterville, Mount.	2,700	50	48	126	22	Admir. chart
*Dupont peak.	5,294	56	54	132	31	Coast Surv.
Dyer, Mount.	6,100	51	00	125	38	Admir. chart
Eagle crag .	5,705	56	55	131	42	Coast Surv.
Eagle lake	3,468	51	53	124	25	C. P. R.
Eagle lake .	2,700	52	00	121	10	C. P. R.
Eagle peak.	9,337	51	16	117	25	Interior
East Barriere lake.	2,250	51	15	119	47	Geol. Surv.
Echo lake.	3,714	51	24	120	30	C. P. R.
Edgar lake.	2,800	59	25	134	10	Geol. Surv.
Edith, Mount.	2,600	50	38	125	47	Admir. chart
Edmund mountain.	6,420	59	46	133	15	Geol. Surv.
Egnell, Mount.	5,400	58	20	131	56	Geol. Surv.
Egremont, Mount	2,800	50	29	125	42	Admir. chart
Eholt.	3,096	49	09	118	36	C. P. R.
*Elbow mountain.	4,233	56	56	131	50	Bdy. Com., Can.
Eldon, Mount .	2,011	50	24	125	31	Admir. chart
Elise mountain.	6,960	49	21	117	11	Geol. Surv.
Eliza dome .	2,819	49	53	127	08	Admir. chart
Eliza Ears, north.	4,107	50	00	127	01	Admir. chart
Eliza Ears, south.	3,885	49	59	127	02	Admir. chart
Eliza, Mount.	1,100	50	57	126	51	Admir. chart
Elk Lake station	235	48	32	123	23	V. & :

BRITISH COLUMBIA

Locality	Elev.	Lat.	Long.	Authority
Elko...		49 18	115 07	C. P. R.
Station	3,082			C. P. R.
E river, 1·4 mile E., high water, 3,021 ; low water, 3,011 ; rail	3,028			C. P. R.
Ellesmere, Mount .	5,800	49 38	123 48	Admir. chart
Elphinstone, Mount...	4,508	49 28	123 33	Admir. chart
Elwin mountain...	2,557	53 56	130 02	Admir. chart
Emily, Mount. .	1,225	50 56	128 53	Admir. chart
*Emmerich, Mount...	6,800	59 13	135 41	Bdy. Com., Can.
*Emmons, Mount...	5,750	59 33	135 23	Bdy. Com., Can.
Enderby .	1,165	50 32	119 08	C. P. R.
Entrance mountain...	1,275	50 28	128 04	Admir. chart
Entwistle, Mount.	4,000	50 35	125 30	Admir. chart
Erickson.	2,107	49 05	116 13	C. P. R.
*Ericsson peak .	4,286	57 09	133 02	Coast Surv.
Erie		49 11	117 20	
Station...	2,343			N. & F. S. R.
Summit, 1·6 mile W...	2,352			N. & F. S. R.
North fork Salmon river, bed, 2,322 ; rail...	2,333			N. & F. S. R.
Erie mountain	5,198	49 13	117 22	Geol. Surv.
Esquimalt...	33	48 25	123 26	E. & N. R.
Estekwalan mountain. .	5,260	50 31	119 39	Geol. Surv.
Estero peak...	6,055	50 28	125 13	Admir. chart
Estevan island...	1,700	53 05	129 38	Admir. chart
Euchiniko lake	3,129	53 07	124 27	C. P. R.
Euptor, Mount...	6,000	49 35	125 38	Admir. chart
Eureka lake...	3,761	51 25	121 00	C. P. R.
Eutiakwetachick lake.	2,978	53 12	125 32	C. P. R.
Evans, Mount (Vancouver island)...	2,500	49 18	124 41	Admir. chart
Evans, Mount (Bute inlet)	6,900	50 58	124 53	Admir. chart
Evening mountain...	1,486	50 13	125 05	Admir. chart
Everard, Mount...	7,000	51 01	125 42	Admir. chart
*Everett peak...	3,748	58 01	133 54	Bdy. Com., Can.
Everingham, Mount...	5,200	50 57	126 16	Admir. chart
Ewing mountain...	5,400	59 52	133 25	Geol. Surv.
Extension...	353	49 07	123 58	E. & N. R.
Face mountain	5,095	50 26	124 44	Admir. chart
	14,500			Malaspina
	14,000			Tebenkof
	13,946			Vasilief
*Fairweather, Mount	14,708	58 55	137 32	Russian chart
	13,864			Tebenkof
	15,500			Coast Surv. '95
	15,287			Bdy. Com., Can.
	15,292			Coast Surv. '00
False Ears...	3,814	50 08	127 09	Admir. chart
Family Humps, east...	2,990	50 03	127 08	Admir. chart
Family Humps, west ...	2,892	50 03	127 09	Admir. chart
Fane, Mount...	1,300	48 33	123 29	Admir. chart
Farnsworth, Mount...	5,730	59 30	133 00	Geol. Surv.
Farrant island...	1,700	53 20	129 26	Admir. chart
Farron...	3,985	49 16	118 08	C. P. R.
Faulkner, Mount. ..	1,600	50 31	126 16	Admir. chart
Fawnie mountains...	6,000	53 10	125 00	Geol. Surv.
Fermanagh, Mount...	1,135	50 56	126 57	Admir. chart
Fernie.		49 30	115 04	
Station...	3,303			C. P. R.
Coal creek, water, 3,277 ; rail ...	3,302			C. P. R.
Fernie mines ...	3,829			C. P. R.
Fiddick...	128	49 30	114 58	C. P. R.
Field		49 06	123 54	E. & N. R.
		51 22	116 30	C. P. R.
Station...	4,064			C. P. R.
Mount Stephen tunnel...	4,344			C. P. R.
'Muskeg' summit, 2·3 miles W...	4,172			C. P. R.
Field, Mount...	8,554	51 26	116 28	Interior
Fife (formerly Sutherland)...	1,978	49 04	118 11	C. P. R.
Finger mountain...	5,000	51 08	126 37	Admir. chart
Finlay, Mount...	5,200	57 34	126 18	Geol. Surv.

BRITISH COLUMBIA

Locality	Elev.	Lat.		Long.		Authority
		°	′	°	′	
Fire-pan mountains............................	7,000-8,000	55	30	126	30	Geol. Surv.
Fish lake..	3,462	50	02	117	11	K. & S. R.
Fisher mountain.........................	9,245	49	30	115	29	Geol. Surv.
Fisherman........................	2,241	49	06	118	20	C. P. R.
Flat mountain	6,000	51	13	126	45	Admir. chart
Flat Top mountain	7,400	53	05	121	00	Geol. Surv.
Flora peak	1,950	50	40	125	46	Admir. chart
Florence, Mount	1,515	50	57	126	55	Admir. chart
Flores island....	3,000	49	18	126	10	Admir. chart
*Foote peak	5,176	57	04	132	45	Coast Surv.
Forbes, Mount	2,000	50	34	125	53	Admir. chart
Forge mountain	5,500	50	34	121	00	Geol. Surv.
Forks mountain	3,600	52	36	121	45	Geol. Surv.
Fortress lake..............	4,350	52	23	117	55	Coleman
Foster, Mount	3,800	50	45	126	01	Admir. chart
Fountain peak	5,640	50	43	121	53	Ge l. Surv.
Fountain ridge	5,820	50	40	121	51	Geol. Surv.
Fox, Mount	10,448	51	09	117	25	Interior
Francis, Mount (Knight inlet)...........	4,800	50	48	125	56	Admir. chart
Francis, Mount (Portland canal)..	5,041	55	08	130	49	Admir. chart
Francois lake..........	2,375	54	00	125	30	Geol. Surv.
Franklin range..............	4,680	50	28	126	40	Admir. chart
*Franklyn peak	4,314	57	08	133	03	Coast Surv.
Franklyn range....	3,290	50	28	125	41	Admir. chart
Fraser	2,760	59	44	135	00	W. P. & Y. R.
Fraser lake.............	2,225	54	03	124	40	Geol. Surv.
Frederic, Mount	5,000	50	43	126	04	Admir. chart
Frederick, Mount (Barkley sound).. ...	2,472	48	59	125	31	Admir. chart
Frederick, Mount (Johnstone strait).	2,655	50	27	125	44	Admir. chart
Frederick William, Mount...	6,137	50	06	123	52	Admir. chart
Freshfield, Mount	12,000	51	40	116	56	Collie
Freya, Mount	8,362	49	42	117	49	Geol. Surv.
Furrer	1,161	50	40	120	09	C. P. R.
Galloway..	2,874	49	22	115	14	C. P. R.
Gambier island........	3,176	49	31	123	26	Admir. chart
Gardner, Mount....	6,300	50	35	125	15	Admir. chart
Gardner, Mount	2,470	49	23	123	24	Admir. chart
*Garfield peak ...	3,920	57	16	133	10	Coast Surv.
Garibaldi peaks, east ...	4,458	50	12	127	13	Admir. chart
Garibaldi peaks, west ...	3,685	50	12	127	15	Admir. chart
Gastineau mountains ...	5,740	50	24	125	46	Admir. chart
Gatcho lake...........	3,527	52	57	125	46	C. P. R.
Geikie, Mount................	11,000	52	43	118	22	Geol. Surv.
Gem hill.	5,490	49	18	117	55	Geol. Surv.
Genelle	1,405	49	12	117	42	C. P. R.
Geneste Cone.............	1,400	50	30	125	53	Admir. chart
Genevieve, Mount	1,800	49	42	126	53	Admir. chart
Geoffrey, Mount.......	1,090	49	31	124	43	Admir. chart
George, Mount	3,700	50	28	125	46	Admir. chart
George mountains.......	3,225	55	33	129	50	Admir. chart
Germansen,.	2,550	55	47	124	40	Geol. Surv.
Gifford peninsula........	1,000	50	05	124	45	Admir. chart
Gil, Mount	3,000	53	16	129	13	Admir. chart
Gilpin.	1,670	49	01	118	19	C. P. R.
Gilson, Mount.............	5,140	50	53	125	43	Admir. chart
Gimli, Mount........	9,215	49	46	117	38	Geol. Surv.
Giscome portage.........	2,100	54	10	122	35	Geol. Surv.
Glacier		51	16	117	30	
Station......	4,093					C. P. R.
Summit of Rogers pass, ground, 4,361 ; rail.	4,351					C. P. R.
Illecillawaet river, 3·3 miles W, bed, 3,689 ; rail....	3,744					C. P. R.
Glacier peak	4,120	51	51	125	33	Admir. chart
Gladsheim mountain	9,275	49	47	117	36	Geol. Surv.
Gladstone, Mount.....	4,800	56	00	129	51	Admir. chart
Gladwin	765	50	51	121	30	C. P. R.
Gladys lake.....	2,915	59	54	133	00	Geol. Surv.
*Glave, Mount.............	6,150	59	36	136	19	Coast Surv.
Glenogle..........	2,991	51	17	116		C. P. R.

BRITISH COLUMBIA

Locality	Elev.	Lat.	Long.	Authority
Glossy mountain	6,150	50 37	121 00	Geol. Surv.
Gloucester, Mount	3,228	50 17	124 53	Admir. chart
Gnawed mountain	5,720	50 25	120 59	Geol. Surv.
Goatfell	2,007	49 08	116 10	C. P. R.
Goat peak	8,200	51 52	117 . '	Collie
Goat Falls Cañon tank	2,119	49 03	116 2,	C. P. R.
Golden		51 19	116 20	
Station	2,580			C. P. R.
Columbia river, water	2,567			C. P. R.
Golden Ears, Mount	5,560	49 22	122 31	Admir. Surv.
Goldstream	283	48 27	123 32	E. & N. R.
Good, Mount	4,000	48 53	124 10	Admir. chart
Goodsir, Mount	11,400	51 12	116 25	Outram
Gosset, Mount	2,300	50 36	126 04	Admir. chart
Grahame, Fort water in Finlay river	2,215	56 36	124 43	Geol. Surv.
Grand Forks		49 02	118 27	
Station	1,746			C. P. R.
Kettle river, low water, 1,670; rail	1,689			C. P. R.
Granite		49 30	117 22	
Station	1,768			C. P. R.
Kootenay river, water, 1,754; rail	1,769			C. P. R.
Granite mountain	6,665	49 06	117 51	Geol. Surv.
Granite peak	6,653	50 44	124 59	Admir. chart
Grant hill	1,240	49 12	122 30	Admir. chart
Grant peak	5,291	57 12	133 04	Coast Surv.
Grassy mountain	6,670	49 18	117 29	Geol. Surv.
Grazebrooke, Mount	5,420	50 21	124 39	Admir. chart
Green lake, No. 1, high water, 2,072; low water	2,070	50 05	123 05	C. P. R.
Green lake, No. 2, high water, 2,092; low water	2,090	50 06	123 04	C. P. R.
Green lake, No. 3, high water, 2,105; low water	2,103	50 09	123 03	C. P. R.
Green lake, No. 4, high water, 2,086; low water	2,084	50 11	123 00	C. P. R.
Green, Mount	8,860	51 14	117 35	Interior
Green mountain	860	50 07	125 00	Admir. chart
Green berry, Mount	6,000	53 05	121 30	Geol. Surv.
Greenwood	2,464	49 05	118 44	C. P. R.
Grey, Mount	4,768	49 00	124 41	Admir. chart
Grey Wolf mountain	7,440	49 58	117 41	Geol. Surv.
Griffin Lake station		50 58	118 31	
Station	1,511			C. P. R.
Eagle river, 2·3 miles E., water, 1,618; rail	1,626			C. P. R.
Griffin mountain	6,850	50 54	118 40	Geol. Surv.
Griffin, Mount	1,410	54 32	130 29	Admir. chart
Grohman mountain	7,550	49 37	117 19	Geol. Surv.
Guanton mountains	5,163	55 45	129 50	Admir. chart
Guardian mountain	5,650	59 34	132 45	Geol. Surv.
Guy mountain	3,500	52 27	121 57	Geol. Surv.
Gwendoline hills	840	50 02	124 18	Admir. chart
Habel, Mount	10,600	51 35	116 36	Outram
Hache, Lake la	2,679	51 50	121 35	C. P. R.
Halcro peak	5,860	50 56	133 44	Geol. Surv.
Half-moon lake	3,854	51 25	120 50	C. P. R.
Hall	2,828	49 22	117 14	N. & F. S. R
Hallowell, Mount	3,980	49 42	123 57	Admir. chart
Hammond	27	49 12	122 39	C. P. R.
Hand, Mount	1,164	52 11	128 14	Admir. chart
Haney	25	49 13	122 35	C. P. R.
Hankin, Mount	2,279	49 11	124 48	Admir. chart
Hankin range	4,000	50 25	126 55	Admir. chart
Hardy, Mount	2,610	50 30	126 04	Admir. chart
Harewood Mount	2,600	50 28	125 45	Admir. chart
Harold, Mount	1,200	50 32	125 39	Admir. chart
Harrison	46	49 14	121 57	C. P. R.
Harrison lake, high water, 42; low water	30	49 30	121 50	C. P. R.
Hartford junction	4,300	49 05	118 35	C. P. R.
Harvey, Mount	5,865	49 28	123 11	Admir. chart
Hastings	29	49 17	123 00	C. P. R.
Hastings hill	1,221	50 31	125 45	Admir. chart
Hawes, Mount	1,400	50 52	126 32	Admir. chart

BRITISH COLUMBIA

Locality	Elev.	Lat.	Long.	Authority
Hawkins, Mount	4,740	50 37	126 08	Admir. chart
Hayes, Mount (Bute inlet)	5,580	50 21	124 56	Admir. chart
Hayes, Mount (Vancouver island)	1,550	49 01	123 54	Admir. chart
Hazelton	725	55 14	127 41	C. P. R.
Heart mountain	5,100	50 30	132 05	Geol. Surv.
Hecate, Mount	3,440	49 01	124 58	Admir. chart
Hector	5,297	51 20	116 20	C. P. R.
Heimdal mountain	8,215	49 41	117 40	Geol. Surv.
Heinze mountain	4,950	49 00	117 40	Geol. Surv.
Hela, Mount	8,915	49 54	117 35	Geol. Surv.
Helena, Mount	5,451	50 13	123 52	Admir. chart
*Hendrickson, Mount	4,550	59 40	130 28	Bdy. Com., Can.
Henry, Mount	3,000	50 44	125 29	Admir. chart
*Henry peak	3,384	57 08	132 56	Coast Surv.
Herbert, Mount	4,200	49 53	126 30	Admir. chart
Hermit mountain	10,194	51 20	117 33	Interior
Hilda, Mount	8,630	49 48	117 55	Geol. Surv.
Hills (formerly Slocan Lake station)	1,785	50 05	117 28	C. P. R.
Hills, Mount	5,050	50 57	125 39	Admir. chart
Hogem pass	3,438	55 30	125 54	Geol. Surv.
Holdsworth, Mount	3,040	50 31	126 56	Admir. chart
Hooker, Mount (Athabaska pass)	10,505	52 28	118 18	Wilcox
Hooker, Mount (Kootenay lake)	8,055	49 40	116 40	Geol. Surv.
Hooper, Mount (Vancouver island)	5,100	49 01	124 28	Admir. chart
Hooper, Mount (Knight inlet)	5,360	50 48	126 08	Admir. chart
*Hoorts, Mount	2,077	59 45	139 31	Coast Surv.
Hope		49 21	121 26	
Station	216			C. P. R.
Fraser river 2·5 miles W. (opposite Huascar bluffs), high water	100			C. P. R.
Hofetzky mountain	5,272	55 38	126 53	Geol. Surv.
Horne lake	357	49 20	124 43	Admir. chart
Horse lake	3,779	51 25	120 55	C. P. R.
Hosmer	3,447	49 35	114 58	C. P. R.
Hotailuh mountain	5,700	58 20	130 15	Geol. Surv.
House cone	2,500	48 40	124 33	Admir. chart
*House peak	6,001	56 56	132 34	Coast Surv.
House mountain	4,118	50 55	124 54	Admir. chart
Howley mountain	6,300	53 00	121 25	Geol. Surv.
Howse pass	4,500	51 46	116 45	C. P. R.
Howse peak	10,900	51 49	116 40	Collie
Hozameen pass	6,277	49 03	121 01	Bdy. Com.
Hudson Hope, water in Peace river	1,522	56 02	121 58	Geol. Surv.
Hughes range	6,000	49 45	115 40	Geol. Surv.
Hulme, Mount	1,065	50 52	127 01	Admir. chart
Hulton, Mount	3,080	50 45	126 13	Admir. chart
Hungabee, Mount	11,305	51 20	116 16	Interior
Hungry peak	8,600	49 45	116 32	Geol. Surv.
Hunter, Mount	8,605	51 17	116 42	Interior
Hunter peak	5,550	55 10	122 10	Geol. Surv.
Hunter range	7,460	50 46	118 50	Geol. Surv.
Huntingdon	47	49 00	122 14	C. P. R.
Huntingdon, Mount	5,048	55 16	129 40	Admir. chart
Huxley island	1,500	52 27	131 23	Geol. Surv.
Ick, Mount	1,365	50 49	126 47	Admir. chart
Ida, Mount	5,200	50 38	119 18	Geol. Surv.
Idaho butte	6,180	59 36	133 17	Geol. Surv.
Ilgachuz mountains	6,000–7,000	52 40	125 20	Geol. Surv.
Illecillewaet		51 11	117 45	
Station	2,719			C. P. R.
Illecillewaet river, 0·6 miles E., bed 2,722 ; rail	2,758			C. P. R.
Indian Point lake	3,200	53 14	121 18	C. P. R.
Ingersoll mountain	7,240	50 03	117 59	Geol. Surv.
Irene mountain	6,750	49 03	116 57	Geol. Surv.
Iron mountain	5,280	50 02	120 46	Geol. Surv.
Irvine, Mount	7,765	49 33	116 57	Geol. Surv.
Isaac lake	3,180	53 15	121 00	Geol. Surv.
*Iskut mountain	4,799	56 41	131 41	Coast Surv.

BRITISH COLUMBIA

Locality	Elev.	Lat.		Long.		Authority
		°	′	°	′	
Island mountain	5,800	53	05	121	41	Geol. Surv.
Itcha mountains	6,000-7,000	52	40	125	00	Geol. Surv.
Jack peak	6,420	59	36	132	42	Geol. Surv.
Jaffray	2,700	49	22	115	18	C. P. R.
James, Mount	2,676	50	43	126	25	Admir. chart
Jauncey, Mount	3,654	55	05	130	05	Admir. chart
Jeffrey, Mount	1,800	48	36	123	33	Admir. chart
Johnson range	5,500	59	26	133	26	Geol. Surv.
Johnson, Mount	6,305	56	03	129	58	Bdy. Com., Can.
Jolliffe, Mount	1,460	50	55	129	58	Admir. chart
Jones, Mount	6,235	50	34	125	19	Admir. chart
Joss mountain	7,000	50	40	118	27	Geol. Surv.
*Kahtate mountain	4,650	56	40	131	42	Bdy. Com., Can.
Kaketsa, Mount	5,000	58	10	132	00	Geol. Surv.
Kamloops	1,160	50	40	120	20	C. P. R.
Kamloops lake, high water, 1,138; low water	1,110	50	45	120	45	C. P. R.
Kanaka (formerly Cisco).		50	07	121	34	
Station	563					C. P. R.
Fraser river, 0·6 mile N., high water (1876), 464; low water (1880), 417; rail.	558					C. P. R.
Kananaskis pass	7,802	50	30	115	18	Wilcox
Kangaroo mountain	4,800	52	42	121	45	Geol. Surv.
*Kaylso, Mount	4,580	56	16	131	36	Bdy. Com., Can.
Karmutzen lake	100	50	25	127	00	Admir. chart
Karmutzen, Mount	5,500	50	21	127	01	Admir. chart
Kaslo	1,752	49	55	116	54	K. & S. R.
*Kates Needle	9,560	57	03	132	02	Bdy. Com., Can.
Keating	201	48	33	123	24	V. & S. R.
Keefer		50	01	121	32	
Station	563					C. P. R.
Fraser river, 3·2 miles S., water (June 29, 1881)	354					C. P. R.
Kennedy island	2,705	54	02	130	11	Admir. chart
Kennedy, Mount	6,655	50	50	125	32	Admir. chart
Keremeos	1,300	49	12	119	48	Geol. Surv.
Kerr, Mount	1,115	50	55	127	01	Admir. chart
Kicking Horse lake	5,294	51	26	116	20	C. P. R.
Kimball mountain	6,600	52	53	121	06	Geol. Surv.
Kimberly	3,657	49	11	115	58	C. P. R.
Kingcome mountain	5,600	50	55	126	09	Admir. chart
Kirby and Spence, Mount	9,290	49	03	114	15	Geol. Surv.
Kirkup mountain	8,505	49	07	117	51	Geol. Surv.
Kispiox	795	55	45	127	44	Rys. & Canals
Kitchener	2,436	49	10	116	20	C. P. R.
Kitkargasse village	1,325	55	38	127	35	Geol. Surv.
Kitty Cone	4,200	50	45	125	36	Admir. chart
Klote lake	4,300	55	58	126	12	Carruthers
Kokanee mountain	9,400	49	45	117	08	Geol. Surv.
Koksilah	24	48	46	123	41	E. & N. R.
Kootenay pass, North						
Summit of Rocky mountains	6,750	49	23	114	36	Geol. Surv.
West summit	6,850	49	20	114	55	Geol. Surv.
Kootenay pass, South						
Summit of Rocky mountains	7,100	49	07	114	08	Geol. Surv.
West summit	5,280	48	56	114	42	Geol. Surv.
Kootenay siding	1,778	49	33	117	14	N. & F. S. R.
Kootenay lake		49	30	117	00	
High water (1894)	1,765					C. P. R.
Low water (1898-99)	1,755					C. P. R.
Kootenay Landing	1,769	49	15	116	11	C. P. R.
Kotsine pass	3,600	55	45	126	45	Geol. Surv.
Kualt	1,159	51	44	119	20	C. P. R.
Kuldo	1,280	55	50	127	56	Rys. & Canals
Kuldo, former site	1,385	55	56	127	54	Rys. & Canals
Kuyakuz lake	3,518	53	08	124	40	Geol. Surv.
Kuyakuz or McMillan mountains	5,000	53	10	124	40	Geol. Surv.
Ladybird mountain	7,567	49	25	117	48	Geol. Surv.
Ladysmith (formerly Oyster Harbour)	68	49	02	123	52	E. & N. R.
Lake mountain (Rossland)	5,410	49	02	117	44	Geol. Surv.

BRITISH COLUMBIA

Locality	Elev.	Lat.	Long.	Authority
		° ′	° ′	
Lake mountain (Milbank sound)	2,270	52 21	128 19	Admir. chart
Lance island	1,177	50 49	128 42	Admir. chart
Landalt, Mount	4,250	48 59	124 17	Admir. chart
Lang, Mount	5,390	50 52	125 30	Admir. chart
Langford	248	48 27	123 30	E. & N. R.
Langford lake	213	48 27	123 30	E. & N. R.
*La Perouse, Mount	10,758	58 34	137 05	Bdy. Com., Can.
Larchtree hill	3,964	49 00	119 13	Bdy. Com.
Lardner, Mount	3,465	50 56	126 32	Admir. chart
Lardo or Trout lake	2,400	50 30	117 30	C. P. R.
Larkin		50 22	119 14	
Station	1,319			C. P. R.
Summit, 1·0 mile S	1,356			C. P. R.
Lasca, Mount	7,804	49 50	117 02	Geol. Surv.
Lascelles, Mount	3,400	50 28	125 47	Admir. chart
*Laura, Mount	7,527	56 50	131 43	Coast Surv.
Lazar, Mount	2,190	48 36	123 47	Admir. chart
Leading hill	1,104	49 51	127 01	Admir. chart
Leading peak (Chatham sound)	2,200	54 31	130 23	Admir. chart
Leading peak (Nootka sound)	2,726	49 30	126 28	Admir. chart
Leading peak (Saanich inlet)	1,346	48 29	123 33	Admir. chart
Leanchoil		51 13	116 04	
Station	3,579			C. P. R.
Summit, 2·0 miles W., ground, 3,681 ; rail	3,677			C. P. R.
Le Blanc lake	3,064	52 03	124 12	C. P. R.
Lefroy, Mount	11,080	51 21	116 16	Interior
Lemon Creek station	1,790	49 42	117 29	C. P. R.
Lemon, Mount	1,200	50 53	127 48	Admir. chart
Lewis range	3,800	51 14	127 19	Admir. chart
Lichen mountain	6,850	51 09	119 17	Geol. Surv.
Lillie, Mount	5,280	50 43	125 38	Admir. chart
Lillooet				
Village	840			C. P. R.
Fraser river near mouth of Cayoosh creek, high water, 669 ; low water	661			C. P. R.
Lillooet lake	620	50 15	122 40	Geol. Surv.
Lily lake	3,510	52 56	125 48	C. P. R.
Lina range	5,885	59 30	133 33	Geol. Surv.
Little Shuswap lake, high water	1,146	50 50	119 40	Geol. Surv.
*Lituya, Mount	11,745	58 48	137 27	Bdy. Com., Can.
Llanover, Mount	2,200	50 11	124 50	Admir. chart
Loco	3,260	49 28	115 51	C. P. R.
Log Cabin	2,928	59 47	134 58	W. P. & Y. R.
Loki, Mount	9,120	49 50	116 45	Geol. Surv.
Lombard, Mount	3,000	49 34	126 33	Admir. chart
Lone Cone	2,332	49 13	125 55	Admir. chart
Lone hill	1,000	50 47	126 43	Admir. chart
Long lake	1,275	50 10	119 20	Geol. Surv.
Long lake	3,700	49 12	118 33	C. P. R.
Lookout mountain	4,420	49 04	117 42	Geol. Surv.
Lookout Point	6,600	50 33	121 28	Geol. Surv.
Loon lake (Kamloops district)	2,820	51 06	121 15	Geol. Surv.
Loon lake (Bute Inlet line)	3,088	52 01	124 13	Geol. Surv.
Loop		49 39	114 46	
Siding, middle	4,237			C. P. R.
Tunnel, east portal, ground, 4,402 ; rail	4,251			C. P. R.
Lost mountain	7,002	49 03	117 10	Geol. Surv.
Lowe, Mount	6,320	51 07	125 29	Admir. chart
Lulu Island	8	49 10	123 05	V. & L. I. R.
Lyell island	1,000	52 40	131 45	Geol. Surv.
Lyell, Mount	12,000	51 56	117 05	Collie
Lytton	695	50 14	121 34	C. P. R.
Lytton mountains	6,520	50 13	121 29	Geol. Surv.
Mabel lake	1,270	50 30	118 45	Geol. Surv.
Mabel mountain	7,560	50 41	118 34	Geol. Surv.
Macdonald, Mount	9,482	51 18	117 28	Interior
Macdonald range	8,000	49 10	114 45	Geol. Surv.
Mackenzie, Mount	7,718	50 57	118 06	Interior

BRITISH COLUMBIA

Locality	Elev.	Lat.	Long.	Authority
		° ′	° ′	
Mackie mountain	7,290	49 15	117 48	Geol. Surv.
Macoun, Mount	9,594	51 12	117 24	Interior
Macpherson, Mount	7,893	.0 56	118 18	Interior
Magin, Mount	1,800	60 45	125 33	Admir. chart
Mahood lake	2,081	51 54	120 30	C. P. R.
Maitland, Mount	4,337	49 09	125 29	Admir. chart
Malde mountain	3,795	49 01	117 48	Geol. Surv.
Mamit lake	3,270	50 23	120 49	Geol. Surv.
Maple Grove	27	49 04	122 14	C. P. R.
Mara	1,156	50 41	119 05	C. P. R.
Marion lake	6,257	51 15	117 31	Interior
Mark, Mount	3,080	49 22	124 45	Admir. chart
Marlborough heights	6,114	49 57	123 55	Admir. chart
Marshal, Mount	4,500	55 31	129 59	Admir. chart
Martin mountain	5,190	50 36	119 45	Geol. Surv.
Martley, Mount	6,370	50 48	121 43	Geol. Surv.
Matawadilata range	9,700	56 05	126 00	Carruthers
Material Yard	1,434	49 20	117 45	C. P. R.
Mathison, Mount	4,100	50 51	126 24	Admir. chart
Matthew range	2,365	50 53	126 43	Admir. chart
Matzehtxel.mountain	5,000	54 40	126 17	Geol. Surv.
Mayne island	804	48 51	123 17	Admir. chart
McCallum mountains, east	5,700	59 15	133 33	Geol. Surv.
McCallum mountains, west	5,500	59 15	133 41	Geol. Surv.
McCauley island	1,160	53 40	130 15	Admir. chart
McDonald, Mount	3,235	50 37	125 32	Admir. chart
McDonald peak	1,430	48 26	123 35	Admir. chart
McGillivray	4,158	49 38	115 47	C. P. R.
McGrath, Mount (Chatham sound)	2,220	54 09	130 15	Admir. chart
* McGrath, Mount (Stikine river)	6,179	56 46	131 31	Const Surv.
McGregor mountain	7,320	49 21	116 51	Geol. Surv.
McGuigan	3,515	50 01	117 15	K. & S. R.
McIntosh mountain	5,000	59 50	133 25	Geol. Surv.
McLean, Mount (Kamloops district)	7,600	50 43	122 06	Geol. Surv.
McLean, Mount (Cariboo district)	6,385	59 41	133 02	Geol. Surv.
McLeod lake	2,250	54 55	123 00	Geol. Surv.
McLeod, Mount	6,300	58 35	130 10	Geol. Surv.
McMaster mountain	5,720	59 20	133 10	Geol. Surv.
McMillan or Kuyakuz mountains	5,000	53 10	124 40	Geol. Surv.
McNeil, Mount	4,300	54 35	130 14	Admir. chart
Meadow	2,328	49 11	117 23	N. & F. S. R.
Meadows	2,939	59 40	135 04	W. P. & Y. R.
Meikle, Mount	9,000	56 09	125 35	Carruthers
Menzies mountains, north	3,830	50 14	125 31	Admir. chart
Menzies mountains, south	3,257	50 13	125 30	Admir. chart
Meridian mountain	6,400	52 55	121 33	Geol. Surv.
Metchosin hill	812	48 24	123 32	Admir. chart
Mexicana hill	600	49 04	123 38	Admir. chart
Mica mountain	9,600	52 53	119 30	Geol. Surv.
Michel	3,853	49 42	114 50	C. P. R.
Middle lake	2,645	51 41	124 56	C. P. R.
Midway	1,913	49 01	118 50	C. P. R.
Mill hill	650	48 27	123 29	Admir. chart
Millside	16	49 14	122 51	C. P. R.
Mills mountain	2,396	49 35	126 30	Admir. chart
Mimemooh mountain	6,030	50 11	121 12	Geol. Surv.
Minto, Mount	6,860	59 56	133 53	Geol. Surv.
Mission	27	49 08	122 17	C. P. R.
Mission, Mount	1,310	54 22	130 22	Geol. Surv.
Mista, Mount	8,255	49 41	117 59	Geol. Surv.
Mitchell, Mount	6,800	52 47	120 47	Geol. Surv.
Moberly		50 22	117 01	
Station	2,546			C. P. R.
Blaeberry river, 2⋅8 miles W., bed. 2,553 ; rail	2,572			C. P. R.
Moberly lake	2,050	55 52	121 45	Geol. Surv.
Moberly peak	7,780	51 22	116 55	Interior
Monashee mountain	5,900	50 08	118 27	Geol. Surv.

BRITISH COLUMBIA

Locality	Elev.	Lat.	Long.	Authority
		° ′	° ′	
Monk, Mount	2,540	50 16	124 52	Admir. chart
Monte hills, highest	2,580	50 27	119 57	Geol. Surv.
Moore, Mount	4,960	52 20	126 52	Admir. chart
Moose lake, high water, 3,404; water (Oct. 1875)	3,399	52 57	118 55	C. P. R.
Moriarty, Mount	5,185	49 08	124 28	Admir. chart
Morrissey	3,101	49 20	115 00	C. P. R.
Morton, Mount	1,545	50 55	126 56	Admir. chart
Mother Lode mine	3,450	49 07	118 44	C. P. R.
Mountain	2,385	49 29	117 14	N. & F. S. R.
Mowitch	1,179	51 47	119 08	C. P. R.
Moyelle (now Aldridge)	3,051	49 16	115 50	C. P. R.
Moyie	3,046	49 17	115 50	C. P. R.
Moyie lakes, high water	3,040	49 17	115 50	C. P. R.
Mukwistikwan lake	3,762	51 25	120 58	C. P. R.
Mummery, Mount	12,000	51 36	116 48	Collie
Munro mountains	4,740	59 37	133 35	Geol. Surv.
Murchison, Mount	6,126	49 44	123 14	Admir. chart
Murphy	1,367	49 19	117 44	C. P. R.
Murray, Mount	6,000	53 06	121 31	Geol. Surv.
Murray, Mount	6,880	50 31	121 32	Geol. Surv.
Mussen mountain	5,000	59 14	134 02	Geol. Surv.
Myrtle or Cache lake	3,650	52 12	119 45	C. P. R.
Nacus mountain	7,000	55 14	125 50	Geol. Surv.
Nadina mountain	5,255	54 01	126 53	Geol. Surv.
Nakatl, Mount or Pope's Cradle	4,800	54 30	124 36	Geol. Surv.
Nakusp	1,413	50 14	117 48	C. P. R.
Nanaimo	125	49 10	123 56	E. & N. R.
Natalkuz lake, high water, 2,653; low water	2,647	53 25	125 15	C. P. R.
Naumulten mountain	8,010	49 52	117 57	Geol. Surv.
Nazco lake	2,961	52 40	123 26	C. P. R.
Needle peak, (Finlayson channel)	2,800	52 48	128 42	Admir. chart
Needle peak (Nootka sound)	4,475	49 51	126 33	Admir. chart
Needle peak (Bute inlet)	8,145	50 43	124 48	Admir. chart
Nelson island	1,500	49 43	124 00	Admir. chart
Nelson lake	2,900	59 30	134 12	Geol. Surv.
Nelson, Mount (Selkirk mountains)	10,000	50 22	116 15	Geol. Surv.
Nelson, Mount (Nelson district)	5,800	49 32	117 18	Geol. Surv.
Nelson		49 29	119 17	
Can. Pac. station	1,769			C. P. R.
N. & F. S. station	1,781			N. & F. S. R.
Nelson siding (now Mountain station)	2,385	49 29	117 14	N. & F. S. R.
Netaltzul peak	8,600	55 15	127 00	Geol. Surv.
Newcastle range	2,000	50 25	126 12	Admir. chart
New Denver	1,800	49 59	117 23	C. P. R.
Newell, Mount (Cariboo district)	4,200	52 47	121 57	Geol. Surv.
*Newell, Mount or Parsons peak (Lynn canal)	5,545	59 28	135 15	Bdy. Com., Can.
Newport mountain, north	5,538	55 24	129 55	Admir. chart
Newport mountain, south	5,710	55 22	129 56	Admir. chart
Newton, Mount	800	48 37	123 28	Admir. chart
New Westminster		49 13	122 55	
Station	13			C. P. R.
City reservoir, water (full)	407			Hill
Nicholls, Mount	1,900	48 57	123 51	Admir. chart
Nicoamen plateau	5,340	50 20	121 18	Geol. Surv.
Nicola lake	2,127	50 10	120 30	C. P. R.
Nicola plateau	5,000	50 15	121 35	Geol. Surv.
Nicomen		49 11	122 05	
Station	32			C. P. R.
Fraser river, 0·3 mile E., high water, 29; low water	20			C. P. R.
Nightingale, Mount	1,675	50 57	126 57	Admir. chart
Niles, Mount	9,510	51 30	116 25	Interior
Niord, Mount	8,815	49 59	117 34	Geol. Surv.
Nipple summit	2,876	50 13	124 58	Admir. chart
Nisconlith lake	1,620	50 47	119 46	Geol. Surv.
Noble mountain	2,874	53 39	130 11	Admir. chart
Noel hill	1,000	51 07	127 38	Admir. chart
Noolki lake, high water, 2,380; low water	2,377	53 55	124 10	C. P. R.
Nootka Cone	1,619	49 37	126 40	Admir. chart

BRITISH COLUMBIA

Locality	Elev.	Lat.	Long.	Authority
Norns mountains, east.	7,365	49 30	117 42	Geol. Surv.
Norns mountains, west	8,115	49 30	117 47	Geol. Surv.
North Bend	495	49 51	121 26	C. P. R.
North Fork pass	6,773	49 55	114 43	Geol. Surv.
North Star junction	2,981	49 33	115 45	C. P. R.
North Star mountain.	6,510	49 02	116 52	Geol. Surv.
Nose peak	1,730	50 29	128 00	Admir. chart
Notch Hill station	1,682	50 51	119 26	C. P. R.
Nubble Mountain	1,400	53 49	130 35	Admir. chart
Ogden peak	7,500	52 36	120 52	Geol. Surv.
O'Hara lake	6,480	51 21	116 19	Interior
O. K. mountain	5,090	49 04	117 51	Geol. Surv.
Okanagan lake, high water, 1,135; low water	1,130	50 00	119 20	C. P. R.
Okanagan Landing	1,138	50 14	119 22	C. P. R.
Okanagan mountain	3,740	49 45	119 30	Geol. Surv.
Oke, Mount	1,000	50 56	127 01	Admir. chart
O'Keefe mountain	5,500	50 08	133 14	Geol. Surv.
Oldfield, Mount	2,300	54 17	130 15	Admir. chart
Old Hogem	2,570	55 46	125 25	Geol. Surv.
Olson tank	3,529	49 39	115 55	C. P. R.
Omineca lake	4,106	56 11	126 40	Carruthers
One-island lake	3,759	51 25	120 55	C. P. R.
Ootsabunkut lake	2,740	53 40	126 00	Geol. Surv.
Oro Denoro mine	3,400	49 08	118 32	C. P. R.
Osmington, Mount	4,500	50 36	125 26	Admir. chart
Osoyoos lake, mean	946	49 00	119 27	Bdy. Com.
Ottertail	3,696	51 18	116 04	C. P. R.
Otway, Mount	1,215	50 52	126 56	Admir. chart
Owen, Mount	4,960	50 41	125 26	Admir. chart
Owen, Mount	10,000	51 19	116 25	Interior
Oyster Bay siding	58	49 00	123 49	E. & N. R.
Oyster Harbour station (now Ladysmith)	68	49 02	123 52	E. & N. R.
Ozzard, Mount	2,270	48 58	125 30	Admir. chart
Pachena Cone	2,223	48 52	125 03	Admir. chart
Paddy peak	8,162	49 59	117 08	Geol. Surv.
Paget Cone	2,600	50 27	125 40	Admir. chart
Palliser		51 15	116 12	
Station	3,283			C. P. R.
Kicking Horse river, 1·5 mile E., water, 3,305; rail.	3,318			C. P. R.
Palmer mountain (Laredo channel)	1,000	52 41	128 54	Admir. chart
Palmer mountains, north (Cariboo)	6,500	53 01	121 21	Geol. Surv.
Palmer mountains, south (Cariboo)	6,700	52 59	121 19	Geol. Surv.
Palmerston, Mount	5,000	50 25	126 20	Admir. chart
Pardoe, Mount	5,000	50 10	124 36	Admir. chart
Park mountains	7,540	50 32	118 37	Geol. Surv.
Park siding		49 52	117 38	
Station	1,622			C. P. R.
Slocan river, extreme flood	1,618			C. P. R.
Parke, Mount	804	48 50	123 18	Admir. chart
Parson Bridge	83	48 27	123 27	E. & N. R.
*Parsons peak or Mount Newell	5,545	59 28	135 15	Bdy. Com., Can.
Parsons, Mount	4,554	49 00	124 43	Admir. chart
Parthenope, Mount	1,345	50 58	126 51	Admir. chart
Pasmore, Mount	2,800	51 57	126 33	Admir. chart
*Patterson peak	4,812	56 58	132 41	Bdy. Com., Can.
Pauls peak	3,570	50 42	120 17	Geol. Surv.
Pavilion lake	2,652	50 50	121 44	C. P. R.
Pavilion mountains, north	6,920	50 58	121 41	Geol. Surv.
Pavilion mountains, south	6,800	50 52	121 40	Geol. Surv.
Pavy	2,170	59 55	134 53	W. P. & Y. R.
Pennington	2,172	59 59	134 53	W. P. & Y. R.
Penny	1,260	50 45	120 59	C. P. R.
Perley rock	7,898	51 14	117 24	Interior
Perry ridge	7,020	49 36	117 38	Geol. Surv.
Petlooshkwohap Mountain	9,180	50 13	121 51	Geol. Surv.
Phœnix	4,625	49 06	118 35	C. P. R.
Pike lake	2,500	59 15	133 25	Geol. Surv.
Pinaoos lake	3,120	50 25	119 36	Geol. Surv.

BRITISH COLUMBIA

Locality	Elev.	Lat. °	Lat. '	Long. °	Long. '	Authority
Pine City	2,825	59	36	133	34	Geol. Surv.
Pine pass	2,850	55	24	122	37	Geol. Surv.
Pine ridge	5,900	49	24	117	55	Geol. Surv.
*Pinta, Mount	8,900	59	56	138	48	Bdy. Com., Can.
Pitt, Mount	3,000	53	24	129	29	Admir. chart
Plowden, Mount	4,900	50	58	126	27	Admir. chart
Plumridge, Mount	3,355	50	53	126	31	Admir. chart
Pocock, Mount	2,300	53	26	129	17	Admir. chart
Poett heights	2,753	48	52	124	58	Admir. chart
Poison hill	5,370	51	05	120	25	Geol. Surv.
Pollux, Mount	9,100	51	12	117	29	Interior
Porcupine ridge	6,030	50	59	120	34	Geol. Surv.
Porte d'Enfer, summit	3,790	54	12	124	40	Geol. Surv.
Porteous	3,008	49	07	115	53	C. P. R.
Port Moody	13	49	17	122	51	C. P. R.
Porto Rico	2,587	49	19	117	14	N. & F. S. R.
Poupore	1,544	49	13	117	41	C. P. R.
Powell, Mount (Bute inlet)	6,066	50	28	124	58	Admir. chart
Powell, Mount (Nootka inlet)	3,083	49	36	126	29	Admir. chart
Pratt, Mount	5,330	50	40	125	25	Admir. chart
Prescott island	820	54	05	130	37	Admir. chart
Prescott, Mount	3,000	50	50	126	29	Admir. chart
Prevost, Mount	2,687	48	51	123	46	Admir. chart
Prideaux, Mount	5,060	50	44	125	55	Admir. chart
Prince of Wales range	4,645	50	21	125	52	Admir. chart
Princeton (Allison or Vermilion Forks)	1,885	49	27	120	28	Geol. Surv.
Proctor siding	1,726	49	38	117	32	C. P. R.
Promise island	1,710	52	23	129	16	Admir. chart
Ptarmigan hill	6,331	49	08	120	26	Bdy. Com.
Pukeashun mountain	7,390	51	12	119	14	Geol. Surv.
Puntzee lake	3,121	52	12	124	05	C. P. R.
Purity, Mount	9,060	51	05	117	29	Topham
Pyramid mountain	8,500	49	44	116	15	Geol. Surv.
Pyramid peak	4,070	58	52	136	17	Coast Surv.
Quadra hill	740	48	56	123	30	Admir. chart
Qualcho lake	2,765	53	00	125	52	C. P. R.
Queest mountain, north	6,682	51	03	118	48	Geol. Surv.
Queest mountain, south	7,081	51	02	118	48	Geol. Surv.
Quesnel lake	2,250	52	30	121	00	Geol. Surv.
Quoin or Stoney mountain	1,500	50	49	126	38	Admir. chart
Rampart ridge	8,407	51	13	117	30	Interior
Raza island	3,020	50	18	125	02	Admir. chart
Read, Mount (Kingcome inlet)	4,820	50	46	126	16	Admir. chart
Read, Mount (Loughborough inlet)	4,420	50	40	125	28	Admir. chart
Reco mountain	8,560	50	00	117	11	Geol. Surv.
Redgrave	2,540	51	31	117	16	C. P. R.
Red mountain (Nelson)	7,250	49	24	117	20	Geol. Surv.
Red mountain (Rossland)	5,150	49	05	117	48	Geol. Surv.
Red mountain (Vancouver island)	2,348	50	06	127	29	Admir. chart
Red Top mountain	3,000	52	47	131	58	Geol. Surv.
Remarkable Cone	2,300	52	09	127	52	Admir. chart
Revelstoke		51	00	118	11	
Station	1,503					C. P. R.
Columbia river, extreme high water, 1,450; low water, 1,432; rail	1,481					C. P. R.
*Reverdy mountains	5,117	56	00	130	00	Coast Surv.
Rinda mountain	8,275	49	44	117	43	Geol. Surv.
*Ripinski, Mount	4,090	59	21	135	20	Coast Surv.
*Robertson, Mount	7,150	56	54	131	37	Coast Surv.
Robinson, Mount	2,100	51	11	127	36	Admir. chart
Robson	1,414	49	20	117	41	C. P. R.
Robson peak	13,700	53	06	119	10	Geol. Surv.
Robson ridge, east end	5,220	49	19	117	44	Geol. Surv.
Robson ridge, west end	5,290	49	19	117	53	Geol. Surv.
Rochers Déboulés range	6,680	55	11	127	40	Geol. Surv.
Roches, Lac des	3,729	51	24	120	40	C. P. R.
Rodell, Mount	7,280	51	01	125	28	Admir. chart
Roderick, Mount	4,356	49	43	123	17	Admir. chart

BRITISH COLUMBIA

Locality	Elev.	Lat.		Long.		Authority
		°	′	°	′	
Rodney, Mount	7,883	50	51	124	49	Admir. chart
Rogers, Mount	10,528	51	20	117	32	Interior
Rogers Pass station		51	19	117	31	
Old station	4,204					C. P. R.
New station	4,309					C. P. R.
Rogers pass	4,351	51	18	117	30	C. P. R.
Rosa, Mount	2,553	49	48	126	55	Admir. chart
Rosebery	1,795	50	02	117	26	C. P. R.
Rossland		49	05	117	47	
Can. Pac. station	3,461					C. P. R.
Union avenue	2,975					C. P. R.
Le Roi mine, end of track	3,742					C. P. R.
Crown Point	2,638					C. P. R.
Red Mt. station	3,488					R. M. R.
Ross peak	7,718	51	14	117	34	Interior
Ross Peak station	3,456	51	15	117	36	C. P. R.
Round Saddle	2,310	49	26	126	24	Admir. chart
Round Top mountain	6,700	52	53	121	22	Geol. Surv.
Royal Oak	115	48	30	123	22	V. & S. R.
Royston, Mount	2,625	50	27	125	51	Admir. chart
Ruby Creek station	102	49	20	121	36	C. P. R.
Rugged peak	4,000	49	54	125	33	Admir. chart
*Ruhamah, Mount	8,445	50	50	139	03	Coast Surv.
Ruskin	30	49	10	122	25	C. P. R.
Russell	43	48	26	123	23	E. & N. R.
Russell, Mount	9,321	51	22	116	40	Interior
Rykert mountain	6,520	49	02	116	44	Geol. Surv.
Saanich hill	669	48	31	123	36	Admir. chart
Saanichton	170	48	35	123	24	V. & S. R.
Saddle hill	550	48	42	123	28	Admir. chart
*Saddle mountain (Lynn canal)	5,890	58	58	135	33	Bdy. Com., Can.
Saddle mountain (Sooke)	1,937	48	26	123	39	Admir. chart
Saddle mountain (Quatsino sound)	1,396	50	30	127	55	Admir. chart
Sail Cone	2,290	50	42	125	56	Admir. chart
St. John, Fort—water in Peace river	1,462	56	11	120	53	Geol. Surv.
St. John, Mount	2,680	50	43	125	33	Admir. chart
St. Mary lake	3,125	49	38	116	11	Geol. Surv.
St. Pauls Dome	1,885	50	03	127	25	Admir. chart
Salmo	2,181	49	12	117	16	N. & F. S. R.
Salmon Arm	1,158	50	42	119	16	C. P. R.
San Christoval mountains	4,000	52	30	131	52	Geol. Surv.
Sandon		49	57	117	14	
K. & S. station	3,516					K. & S. R.
Can. Pac. station	3,488					C. P. R.
San Josef, Mount	4,200	50	30	125	16	Admir. chart
Sandy mountain	7,020	51	05	119	13	Geol. Surv.
Sanford, Mount	6,025	59	29	132	48	Geol. Surv.
Sangrida, Mount	8,270	49	39	117	59	Geol. Surv.
Sapperton	15	49	14	122	52	C. P. R.
Savona	1,164	50	45	120	50	C. P. R.
Savona mountain	4,820	50	42	120	49	Geol. Surv.
Sawyer, Mount	4,140	50	39	125	52	Admir. chart
Sayward	1,613	49	03	117	35	N. & F. S. R.
Scaife, Mount	2,253	50	31	125	53	Admir. chart
Scalping Knife mountain	6,915	50	05	117	51	Geol. Surv.
Scarped mountains	5,860	50	19	121	29	Geol. Surv.
Scriven, Mount	4,290	50	38	125	53	Admir. chart
Section mountain	5,320	59	26	133	57	Geol. Surv.
Selkirk Mount	7,700	50	55	115	59	Geol. Surv.
Selwyn, Mount	6,220	56	00	123	45	Geol. Surv.
Semlin	1,156	50	46	121	08	C. P. R.
Sentinel mountain	5,470	49	22	117	36	Geol. Surv.
Serjeant, Mount	2,222	49	37	126	28	Admir. chart
Seton lake, high water, 800; low water	745	50	39	122	00	C. P. R.
Sharp peak	2,975	50	13	127	44	Admir. chart
Sharp Snow summit	3,787	50	06	127	02	Admir. chart
Shawnigan lake, high water, 380; low water	376	48	38	123	40	E. & N. R.
Shawnigan Lake station	383	48	39	123	38	E. & N. R.

BRITISH COLUMBIA

Locality	Elev.	Lat.	Long.	Authority
		° ′	′ ′	
Sheep mountain	8,100	59 08	129 06	Geol. Surv.
Sheep Creek station.	2,223	49 00	117 49	R. M. R.
Sheepshanks hill.	960	48 34	123 34	Admir. chart
Shepherd, Mount (Texada island).	2,900	49 32	124 12	Admir. chart
Shepherd, Mount (Vancouver island).	1,908	48 26	123 39	Admir. chart
Sherbrooke lake.	8,310	51 27	116 23	Interior
Shields.	2,025	49 20	117 55	C. P. R.
Shields, Mount.	2,020	53 50	130 17	Admir. chart
Shields mountain.	6,040	49 21	118 00	Geol. Surv.
Ship peak.	3,197	49 59	127 09	Admir. chart
Shushartie saddle	1,900	50 49	127 52	Admir. chart
Shuswap station.	1,154	50 47	119 42	C. P. R.
Shuswap lake, high water, 1,132; low water.	1,136	51 00	119 00	C. P. R.
Shute, Mount.	2,200	50 32	125 51	Admir. chart
Sicamous.	1,156	50 50	119 00	C. P. R.
Siding No. 1 (now Genelle).	1,405	49 12	117 42	C. P. R.
Siding No. 2 (now Poupore).	1,544	49 13	117 41	C. P. R.
Siffleur lake.	3,753	53 45	127 05	C. P. R.
Sifton, Mount	9,643	51 20	117 35	Interior
Silica.	2,958	49 04	117 50	R. M. R.
Silver hills.	5,900	50 22	118 39	Geol. Surv.
Silver mountain	7,795	50 04	117 41	Geol. Surv.
Silver ridge.	7,026	49 57	117 14	Geol. Surv.
Silverlock hills.	1,900	50 58	127 04	Admir. chart
Silverton	1,799	49 57	117 21	C. P. R.
Silwhoiskun mountain.	6,030	51 00	120 30	Geol. Surv.
Simmonds, Mount.	1,125	50 52	126 57	Admir. chart
Simpson pass	6,550	51 02	115 41	Geol. Surv.
*Simpson peak	5,271	56 51	132 31	Admir. chart
Sinclair pass.	4,662	50 41	116 00	Geol. Surv.
Sirdar junction	1,772	49 16	116 08	C. P. R.
Sirdar station	1,802	49 15	116 07	C. P. R.
Sir Donald, Mount.	10,808	51 15	117 25	Interior
Sitchiada, Mount.	7,500	56 06	126 33	Carruthers
Siwash mountain	7,700	49 21	117 12	Geol. Surv.
Six-mile Creek station.	2,617	51 29	117 29	C. P. R.
Skagit ford.	1,636	49 08	121 11	Bdy. Com.
Skihist mountain	9,600	50 11	121 47	Geol. Surv.
Skoatl point.	5,450	51 09	120 25	Geol. Surv.
Skwilkwakwil mountain	5,660	50 20	121 05	Geol. Surv.
Slate mountain	2,607	50 05	127 06	Admir. chart
Slide mountain	3,800	52 39	121 51	Geol. Surv.
Slocan junction.	1,637	49 27	117 31	C. P. R.
Slocan lake, high water (1894), 1,773; low water (1897).	1,761	50 00	117 25	C. P. R.
Slocan Lake (now Hills) station	1,785	50 05	117 28	C. P. R.
Slocan ridge.	6,650	49 32	117 33	Geol. Surv.
Slocan station.	1,777	49 46	117 27	C. P. R.
Sloko mountain	5,560	59 09	133 45	Geol. Surv.
Smallpeace, Mount	5,230	59 30	132 55	Geol. Surv.
Smelter junction.	1,565	49 06	117 42	C. P. R.
Smith, Mount.	7,544	50 50	125 03	Admir. chart
Smyth Cone.	3,890	50 53	126 08	Admir. chart
*Snow Cap mountain	4,950	56 54	131 52	Bdy. Com., Can.
Snowdrift mountain	4,000	51 07	126 37	Admir. chart
Snow mountain	3,000	50 33	126 09	Admir. chart
Snow Saddle mountain	4,151	50 16	127 15	Admir. chart
*Snow Tower.	6,550	58 10	133 24	Bdy. Com., Can.
Snowdon mountain	6,895	59 43	132 30	Geol. Surv.
Solitary hill.	1,166	49 09	122 00	Admir. chart
Solitary, Mount.	1,718	50 26	125 40	Admir. chart
Solway hill.	4,900	49 12	117 47	Geol. Surv.
Somenos.	83	48 50	123 44	E. & N. R.
Sophia, Mount.	4,200	50 53	126 21	Admir. chart
Sophia range.	2,500	49 48	126 52	Admir. chart
Soues, Mount	7,000	51 04	121 43	Geol. Surv.
South Fork	2,291	49 57	117 00	K. & S. R.
South or Allison pass	5,808	49 22	120 47	C. P. R.
South Wellington	123	49 07	123 54	E. & N. R.

BRITISH COLUMBIA

Locality	Elev.	Lat.		Long.		Authority
		°	′	°	′	
Sovereign mountain..	5,400	52	58	121	56	Geol. Surv.
Spaist mountain	5,780	50	23	121	04	Geol. Surv.
Sparks Cone.......	1,155	50	29	125	58	Admir. chart
Sparshot, Mount...	2,740	50	39	125	30	Admir. chart
Sparwood	3,637	49	43	114	54	C. P. R.
Spatsum......		50	33	121	17	
Station......	859					C. P. R.
Thompson river, 0·7 mile E., high water, 865 : low water	856					C. P. R.
Spencer, Mount (Johnstone strait)......	2,345	50	32	125	50	Admir. chart
Spencer, Mount (Vancouver island). ...	4,718	49	04	124	39	Admir. chart
Spencer, Mount (Jervis inlet)......................	6,214	50	00	123	51	Admir. chart
Spence Bridge station................... .	776	50	25	121	21	C. P. R.
Sphinx mountain......................	8,370	49	38	116	39	Geol. Surv.
Spicer island...................	827	53	45	130	22	Admir. chart
Spire peaks...............	3,160	50	53	125	40	Admir. chart
Spokane mountain....	6,405	49	05	117	51	Geol. Surv.
Spooner, Mount.....	1,650	50	07	124	42	Admir. chart
Spruce mountain...	5,405	50	35	133	25	Geol. Surv.
Spuzzum............		49	41	121	24	
Station	401					C. P. R.
Spuzzum river, 1·4 mile S., high water, 247; low water, 242; rail....... :	373					C. P. R.
Squaakim lake	52	49	13	122	00	C. P. R.
Squilax	1,322	50	52	119	35	C. P. R.
Stamp, Mount ··.....	4,150	50	43	125	59	Admir. chart
Stanley mountain	7,925	49	32	117	56	Geol. Surv.
Stanton peak	1,609	48	58	123	49	Admir. chart
Steele........	2,687	49	32	115	38	C. P. R.
Steeple mountain .. .	7,590	49	15	116	48	Geol. Surv.
Steeple mountain.....	4,402	49	20	125	45	Admir. chart
Stein mountain.......	8,900	50	26	121	47	Geol. Surv.
Stephen....	5,321	51	27	116	18	C. P. R.
Stephen, Mount. ...	10,523	51	24	116	26	Interior
Stephens, Mount	5,665	50	58	126	41	Admir. chart
Stephens island	1,340	54	07	130	37	Admir. chart
Stevenson, Mount....	7,100	52	37	121	10	Geol. Surv.
Steveston	6	49	08	123	11	V. & L. I. R.
Stewart	43	48	29	123	26	E. & N. R.
Stokes, Mount....	3,652	50	27	125	11	Admir. chart
Stone Nipples, north......	3,360	49	59	127	11	Admir. chart
Stone Nipples, south. ...	3,464	49	58	127	11	Admir. chart
Stony mountain.	1,409	48	46	123	16	Admir. chart
Stopford, Mount	940	50	51	126	51	Admir. chart
Stovel peak	5,750	59	54	134	07	Geol. Surv.
Strahan, Mount..	5,289	49	25	123	12	Admir. chart
Stripe mountain.....	2,020	52	26	128	25	Admir. chart
Striped mountain	2,700	49	22	120	05	Geol. Surv.
Stuart island	800	50	23	125	10	Admir. chart
Stuart lake.......... ..	2,200	54	30	125	00	Geol. Surv.
Stump lake.......	2,450	50	22	120	22	Geol. Surv.
Sugar lake	1,980	50	25	118	30	Geol. Surv.
Sugar Loaf, Mount	9,250	51	01	117	23	Topham
*Sugar Loaf mountain.......	5,259	56	50	131	41	Coast Surv.
Sullivan....	1,376	49	11	117	42	C. P. R.
Sumas mountain	3,220	49	03	122	02	Admir. chart
*Sumdum, Mount	6,690	57	48	133	25	Bdy. Com., Can.
Summerfield, Mount...	3,750	50	37	126	08	Admir. chart
Summit lake..........	4,432	49	38	114	42	C. P. R.
Summit lake, high water, 2491; low water.....	2,489	50	05	117	28	C. P. R.
Summit lake, high water, 1807; low water.....	1,804	50	58	118	20	C. P. R.
Summit lake.......... ..	3,509	49	38	121	00	C. P. R.
Summit lake, high water, 1582; low water	1,580	50	27	122	38	C. P. R.
Summit lake...............	2,300	54	15	122	40	Geol. Surv.
Summit siding.....	3,036	49	24	117	12	N. & F. S. R.
Summit siding.	2,500	50	09	117	37	C. P. R.
Summit station......	916	48	34	135	35	E. & N. R.
Sunday peak	6,245	59	45	134	05	Geol. Surv.
Superb mountain	8,000	50	48	124	50	Admir. chart

3

BRITISH COLUMBIA

Locality	Elev.	Lat.	Long.	Authority
		° ′	° ′	
Surprise lake	3,030	59 40	133 15	Geol. Surv.
Suskwa pass	3,772	55 20	126 50	Geol. Surv.
Swampfly peak	4,430	51 03	125 40	Admir. chart
Swan lake	1,285	59 19	119 15	Geol. Surv.
Swansea		49 25	115 51	
Station	3,190			C. P. R.
Summit, 3·9 miles N., ground, 3267 ; rail	3,260			C. P. R.
Swiss peaks, middle peak	10,515	51 21	117 32	Interior
Sydney	22	48 39	123 25	V. & S. R.
Sylvester peak	7,000	59 10	129 15	Geol. Surv.
Table mountain (Parsnip river)	3,500	55 38	123 30	Geol. Surv.
Table mountain (Vancouver island)	3,080	50 30	127 04	Admir. chart
Table mountain (Queen Charlotte islands)	1,720	53 12	131 55	Geol. Surv.
Tachick lake, high water, 2,350 ; low water	2,347	53 58	124 12	C. P. R.
Tacla lake	2,270	55 30	126 00	Geol. Surv.
Tagish lake	2,161	59 45	134 15	Geol. Surv.
Tahaetkun mountain	6,630	50 16	119 44	Geol. Surv.
Taku mountains	5,955	59 43	133 59	Geol. Surv.
Talus hill	3,070	49 06	117 47	Geol. Surv.
Tamarack mountain	4,300	49 02	117 48	Geol. Surv.
Tappen	1,161	50 46	119 20	C. P. R.
Tatelkuz lake	3,049	53 15	124 45	Geol. Surv.
Tatla lake	3,018	52 00	124 30	C. P. R.
Tatlayoco lake	2,723	51 33	124 27	C. P. R.
Tchetlatchudi lake	3,800	56 22	125 52	Carruthers
Tchuctazi lake	3,750	56 17	125 45	Carruthers
Tehuditcho lake	2,800	56 35	125 08	Carruthers
Tchukwazziti range	7,000-8,500	56 20	126 00	Carruthers
Telegraph Creek, water in Stikine river	540	57 55	131 10	Geol. Surv.
Tenakihi range	7,000-8,500	56 13	125 45	Carruthers
'Tent mountain	7,100	57 55	133 08	Admir. chart
Terminal peak (formerly Greens peak)	9,719	51 15	117 25	Interior
Teslin lake	2,600	60 00	132 05	Geol. Surv.
Tetachuck lake	2,770	53 20	125 59	Geol. Surv.
'The Dome'	8,892	51 12	117 20	Interior
'The Steeples'	9,000	49 33	115 26	Geol. Surv.
Thomas, Mount	1,650	50 34	126 19	Admir. chart
Thompson	678	50 16	121 26	C. P. R.
Thompson, Mount (Yellowhead pass)	8,790	52 42	119 08	Geol. Surv.
Thompson, Mount (North Saskatchewan)	10,700	51 37	116 29	Collie
Thompson pass	6,800	52 00	117 20	Collie
Thornton Bate, Mount	2,200	50 58	127 00	Admir. chart
Three Finger peak	5,610	50 51	125 31	Admir. chart
Three Forks	2,594	50 00	117 17	C. P. R.
Three Sisters mountain	6,900	52 42	121 02	Geol. Surv.
Three-valley lake	1,631	50 56	118 28	C. P. R.
Thrums	1,532	49 21	117 34	C. P. R.
Thumb peak	2,500	54 33	130 54	Admir. chart
Tiger	2,483	49 04	117 45	C. P. R.
Tilley, Mount	8,064	50 57	118 05	Interior
Toad mountain	7,334	49 25	117 19	Geol. Surv.
Toba, Mount	2,900	50 20	125 05	Admir. chart
Tochty	2,964	49 12	115 59	C. P. R.
Tod, Mount	7,230	50 53	119 55	Geol. Surv.
Tom, Mount	500	53 07	121 46	C. P. R.
*Tomlinson, Mount	3,265	55 02	129 57	Admir. chart
Tournay, Mount	5,542	55 27	129 57	Admir. chart
Trachyte hills	5,200	50 48	121 29	Geol. Surv.
Trail	1,364	49 00	117 42	C. P. R.
Tranquille	1,142	50 42	120 30	C. P. R.
Tranquille plateau	5,040	50 52	121 34	Geol. Surv.
Treble mountain	5,000	55 53	129 50	Admir. chart
Treble, Mount	4,000	50 30	125 21	Admir. chart
Trematon, Mount	5,050	49 28	123 18	Admir. chart
Triangle island	680	50 52	129 05	Admir. chart
Trinity hills	4,756	50 29	118 53	Geol. Surv.
Tripp hill	1,900	50 29	125 56	Admir. chart
Troubridge, Mount	4,360	49 49	124 11	Admir. chart
Trout or Lardo lake	2,400	50 30	117 30	C. P. R.

BRITISH COLUMBIA

Locality	Elev.	Lat.	Long.	Authority
		° '	° '	
Tsacha lake..................	3,292	53 62	124 50	C. P. R.
Tsalkom mountain............	5,000	50 55	119 43	Geol. Surv.
Tsawhus lake................	3,240	53 40	122 58	Geol. Surv.
Tsiltsalt ridge...............	6,300	50 59	121 37	Geol. Surv.
Tsinkut lake, high water.	2,391	53 54	124 00	C. P. R.
Tsitsutl mountains...........	6,000	52 40	125 50	Geol. Surv.
Tucker, Mount...............	3,380	50 25	125 19	Admir. chart
Tunnel......................		49 21	118 06	
Station	3,206			C. P. R.
Highest point in tunnel..	3,214			C. P. R.
Summit of construction 'switch-back' .	3,648			C. P. R.
Tupper, Mount...............	9,063	51 20	117 29	Interior
*Turner, Mount..............	6,600	56 47	131 32	Bdy. Com., Can.
Turret mountain.............	1,782	49 04	125 11	Admir. chart
Twin Butte station...........	1,907	51 02	117 59	C. P. R.
*Twin peaks.................	6,557	56 57	132 32	Coast Surv.
Twin peaks, north............	4,520	50 30	127 17	Admir. chart
Twin peaks, south............	4,630	50 29	127 17	Admir. chart
Twist lake..................	2,523	51 34	125 03	C. P. R.
Two Sisters mountain........	6,800	53 11	121 36	Geol. Surv.
*Tyee, Mount...............	4,690	56 12	131 26	Bdy. Com., Can.
*Unana, Mount..............	6,500	59 43	139 18	Coast Surv.
Union island................	1,484	50 02	127 17	Admir. chart
Upper Columbia or Windermere lake.....	2,700	50 29	115 55	Geol. Surv.
Upper Kootenay or Duncan lake.	1,800	50 29	117 00	Public Works
Urd, Mount.................	8 640	49 52	117 42	Geol. Surv.
Uto peak	9,352	51 16	117 25	Interior
Valley mountain.............	5,000	53 04	121 35	Geol. Surv.
Vancouver..................		49 17	123 05	
Station............	11			C. P. R.
Can. Pac. wharf, west end......	12			C. P. R.
B. M., cut in most easterly step of station.	14·3			C. P. R.
Reservoir, water, 24 ft. on gauge........	250			City Eng.
City hall, floor level.........	38·4			City Eng.
Highest point, south boundary of city, 600 ft. E. of South Granville St.........	248			City Eng.
Extreme high tide (1894)........	+6·9			City Eng.
Extreme low tide............	−9·8			City Eng.
City datum.................	−92·8			City Eng.
Vansittart, Mount............	1,464	50 24	125 47	Admir. chart
Vaux, Mount................	10,741	51 15	116 30	Interior
Verandrye mountain.........	9,234	50 59	116 02	Geol. Surv.
Vermilion pass	5,264	51 13	116 03	Geol. Surv.
Verney, Mount..............	2,225	50 59	126 52	Admir. chart
Vernon.....................	1,255	50 15	119 16	C. P. R.
Victoria....................		48 26	123 22	
E. & N. station...........	29			E. & N. R.
V. & S. station...........	57			V. & S. R.
City Hall, floor level......	55			City Eng.
Parliament buildings, floor level.	25			City Eng.
Government house, stone step.	150			City Eng.
Methodist church, cor. Pandora and Quadra sts., step	75			City Eng.
Highest point within city limits.............	200			City Eng.
City datum..............	−98			City Eng.
Victoria lake...............	1,746	50 57	118 24	C. P. R.
Victoria, Mount (Jervis inlet).......	7,452	50 17	124 00	Admir. chart
Victoria, Mount (Rocky mountains)......	11,150	51 23	116 18	Interior
Victoria peak (Vancouver island).....	7,484	50 04	126 10	Admir. chart
Victory, Mount..............	3,280	50 32	126 05	Admir. chart
View mountain..............	1,600	49 18	124 47	Admir. chart
Village Cone	3,690	51 07	125 33	Admir. chart
Village mountain............	3,500	51 09	126 47	Admir. chart
*Villard, Mount.............	5,000	59 16	135 17	Bdy. Com., Can.
Vingolf mountain............	8,415	50 02	117 33	Geol. Surv.
Violin lake.................	3,090	49 02	117 42	Geol. Surv.
Wabigoon or White Flower lake..	3,746	51 25	120 55	C. P. R.
Wakefield, Mount...........	5,480	50 47	125 33	Admir. chart
Walker, Mount..............	3,646	49 45	126 34	Admir. chart

3½

BRITISH COLUMBIA

Locality	Elev.	Lat.		Long.		Authority
		° '		° '		
Wall mountain........................	7,690	49	11	116	50	Geol. Surv.
Walter, Mount........................	3,360	51	00	126	29	Admir. chart
Waneta..............................	1,358	49	00	117	36	N. & F. S. R.
Wanklyn.............................	3,607	49	05	115	48	C. P. R.
Wapta, Mount	9,160	51	27	116	29	Interior
Wapnttehk mountains.................	10,000	51	30	116	20	Geol. Surv.
Ward, Mount.........................	2,000	50	59	127	00	Admir. chart
Wardner.............................		49	26	115	26	
Station.............................	2,484					C. P. R.
Kootenay river, 1·8 mile S., high water, 2,440 ; low water (1897), 2,433 ; rail.............	2,449					C. P. R.
Warfield station......................	2,018	49	05	117	45	C. P. R.
Washington, Mount...................	5,415	49	45	125	19	Admir. chart
Waterloo lake........................	4,000	52	16	119	58	C. P. R.
Watson, Mount.......................	4,750	50	51	125	43	Admir. chart
Watt, Mount.........................	8,500	52	39	120	49	Geol. Surv.
Weaver, Mount.......................	2,370	50	05	124	42	Admir. chart
Wellington...........................	377	49	14	124	02	E. & N. R.
Wellington, Mount....................	6,155	50	09	123	58	Admir. chart
Wesley, Mount	2,530	49	19	124	39	Admir. chart
Westholme...........................		48	52	123	43	
Station.............................	93					E. & N. R.
Summit, 2·3 miles, S., ground, 156 ; rail......	146					E. & N. R.
Chemainus river, 2·0 miles N., water, 11 ; rail......	28					E. & N. R.
Westminster junction.................	34	49	16	122	47	C. P. R.
West Robson.........................	1,419	49	19	117	41	C. P. R.
Wharnock............................	20	49	10	122	29	C. P. R.
Whatshan lakes......................	2,751	50	00	118	05	Geol. Surv.
Whieldon, Mount.....................	5,520	50	19	124	38	Admir. chart
*Whipple, Mount.....................	6,633	56	37	131	36	Coast Surv.
White Flower or Wabigoon lake.......	3,746	51	25	120	55	C. P. R.
White lake...........................	1,560	50	53	119	15	Geol. Surv.
White Man pass......................	6,807	50	47	115	30	Geol. Surv.
White, Mount........................	9,800	56	12	126	07	Carruthers
White mountain......................	6,000	50	36	121	26	Geol. Surv.
White pass..........................	2,897	59	36	135	07	W. P. & Y. R.
White-water..........................	3,287	50	01	117	07	K. & S. R.
Whymper, Mount.....................	5,120	48	58	124	14	Admir. chart
Whyte, Mount	10,365	51	24	116	17	Interior
Wigwam.............................	1,424	50	49	118	02	C. P. R.
Wiley, Mount........................	5,700	53	08	121	42	Geol. Surv.
William, Mount......................	1,137	50	13	125	09	Admir. chart
Williams lake........................	1,843	52	06	122	05	C. P. R.
Wimbledon mountains................	2,600-3,000	53	20	129	00	Admir. chart
Windermere or Upper Columbia lake..	2,700	50	20	115	55	Geol. Surv.
Winnipeg mine.......................	4,425	49	05	118	34	C. P. R.
Wiwaxy peaks........................	8,230	51	22	116	19	Interior
Woden, Mount.......................	8,871	49	50	117	47	Geol. Surv.
Wolf hill.............................	900	48	28	123	35	Admir. chart
Wood, Mount (Saanich inlet).........	1,600	48	37	123	35	Admir. chart
Wood, Mount (Clayoquot sound)......	4,162	49	22	125	49	Admir. chart
Wood, Mount.........................	5,420	50	55	125	37	Admir. chart
Woods, Mount........................	3,807	55	01	129	51	Admir. chart
Work, Mount.........................	1,440	48	32	123	20	Admir. chart
*Wright, Mount......................	4,945	58	49	136	02	Coast Surv.
Wrottesley, Mount...................	5,836	49	37	123	21	Admir. chart
Wynyard, Mount.....................	1,200	50	52	127	02	Admir. chart
Yahk................................	2,817	49	06	116	06	Geol. Surv.
Yahk mountain.......................	7,000	49	12	115	43	Geol. Surv.
Yale.................................	223	49	33	121	25	C. P. R.
Yanks peak..........................	6,140	52	48	121	29	Geol. Surv.
Yellowhead lake, high water, 3,635 ; water (Sept., 1875).	3,632	52	52	118	30	C. P. R.
Yellowhead pass.....................	3,738	52	54	118	27	C. P. R.
Ymir................................	2,396	49	17	117	12	N. & F. S. R.
Ymir mountain.......................	7,920	49	26	117	07	Geol. Surv.
Yorke, Mount........................	2,165	50	26	125	49	Admir. chart
Young lake...........................	3,000	51	15	121	00	Geol. Surv.
*Young, Mount......................	5,720	58	52	135	34	Bdy. Com., Can.
Young, Mount	4,850	50	46	125	44	Admir. chart
Yuill, Mount.........................	7,300	49	40	117	06	Geol. Surv.

FRANKLIN

Locality	Elev.	Lat.		Long.		Authority
		°	′	°	′	
Arthur, Mount	4,500	81	13	74	30	Admir. chart
Britannia, Mount	1,500	76	55	96	30	Admir. chart
Bruce mountains	4,000-5,000	71	00	72	00	Admir. chart
Bylot island	2,300	73	00	79	00	Admir. chart
Blanche mountain	2,000	76	40	91	30	Admir. chart
Clarendon, Mount	500	73	20	100	30	Admir. chart
Cooper Key mountains	1,800	83	06	70	30	Admir. chart
Challenger mountains	2,000	82	22	81	00	Admir. chart
Cheops, Mount	4,800	82	35	65	30	Admir. chart
Cape Union peak	2,000	82	15	62	20	Admir. chart
Disraeli, Mount	2,500	82	52	67	15	Admir. chart
Eglinton Fiord peak	1,200	70	55	70	30	Admir. chart
Frere, Mount	5,000	82	40	69	30	Admir. chart
Foster, Mount	2,000	82	52	66	45	Admir. chart
Gladstone, Mount	2,500	82	54	67	00	Admir. chart
Grant, Mount	4,900	82	28	65	15	Admir. chart
Giffard, Mount	2,300	82	58	68	15	Admir. chart
Hornby, Mount	2,000	83	00	70	00	Admir. chart
Jeffries range	1,500	76	10	99	30	Admir. chart
Neville, Mount	3,800	81	10	70	30	Admir. chart
North Kent island	2,000	76	40	90	00	Admir. chart
Prince of Wales mountains	2,000	78	30	78	30	Admir. chart
Pullen, Mount	1,700	82	30	62	20	Admir. chart
Pope, Mount	5,000	80	30	76	20	Admir. chart
Rawlinson, Mount	5,000	82	40	68	30	Admir. chart
Sillem island	3,000	71	00	71	30	Admir. chart
Stokes range	1,300	76	20	101	40	Admir. chart
Truter mountains	2,000	75	40	81	00	Admir. chart
Wilmot, Mount	800	76	30	103	00	Admir. chart

KEEWATIN

Locality	Elev.	Lat.	Long.	Authority
		° ′	° ′	
Aberdeen lake	130	64 30	99 30	Geol. Surv.
Attawapiskat or Lansdowne lake	900	52 15	88 00	Geol. Surv.
Baker lake	30	64 10	96 00	Geol. Surv.
Beaverdam lake	710	55 35	98 35	Geol. Surv.
Bluffy lake	1,220	50 47	93 00	Geol. Surv.
Bug lake	1,295	50 52	93 55	Geol. Surv.
Clearwater lake, Lower	1,320	51 12	92 49	Geol. Surv.
Cross lake (Nelson river)	685	54 40	97 40	Geol. Surv.
Fairy lake	1,261	51 42	93 00	Geol. Surv.
Fly lake	1,356	51 00	92 50	Geol. Surv.
Goose lake	1,198	51 48	93 02	Geol. Surv.
Goose lake, Upper	1,262	51 43	92 54	Geol. Surv.
Gull Rock lake	1,146	50 55	93 40	Geol. Surv.
Hair lake	1,178	51 50	93 08	Geol. Surv.
Island lake	900	53 40	94 00	Geol. Surv.
Kaminuriak lake	320	63 00	95 20	Geol. Surv.
Landing lake	580	55 15	97 30	Geol. Surv.
Lansdowne or Attawapiskat lake	905	52 15	88 00	Geol. Surv.
Little Playgreen lake	710	54 00	98 00	C. P. R.
Long-legged lake	1,175	50 45	94 07	Geol. Surv.
Lady Marjorie lake	260	64 10	99 50	Geol. Surv.
Medicine Stone lake, Lower	1,200	50 54	94 06	Geol. Surv.
Medicine Stone lake, Upper	1,210	50 51	94 03	Geol. Surv.
Minto, Mount	1,050	63 50	81 00	Geol. Surv.
Nelson lake	735	55 50	99 00	Geol. Surv.
Net-setting lake	685	55 00	98 40	Geol. Surv.
Paint lake	500	55 28	98 00	Geol. Surv.
Pekangikum lake	1,037	51 48	94 00	Geol. Surv.
Pipestone lake	510	55 35	98 05	Geol. Surv.
Playgreen lake	710	54 15	98 10	C. P. R.
Red lake	1,146	51 05	93 55	Geol. Surv.
St. Joseph, Lake	1,172	51 06	91 00	Geol. Surv.
Sandbar lake	1,035	50 35	93 35	Geol. Surv.
Schultz lake	115	64 45	98 00	Geol. Surv.
Seul, Lac	1,140	50 20	92 30	Geol. Surv.
Shabumeni lake	1,330	51 20	92 45	Geol. Surv.
Shallow lake	1,105	50 45	93 25	Geol. Surv.
Shallow lake, Little	1,106	50 48	93 22	Geol. Surv.
Sipiwesk lake	565	55 00	97 30	Geol. Surv.
Snake lake	1,270	51 00	93 06	Geol. Surv.
Sturgeon lake	1,051	51 46	93 45	Geol. Surv.
Trout lake	1,205	51 10	93 20	Geol. Surv.
Trout lake, Little	1,204	51 02	93 15	Geol. Surv.
Vermilion lake, Little	1,173	51 10	93 45	Geol. Surv.
Waskaiowaka lake	930	56 30	96 40	Geol. Surv.
Wilcox lake	1,030	50 30	93 45	Geol. Surv.
Wintering lake	530	55 25	97 45	Geol. Surv.
Woman lake	1,315	51 10	92 52	Geol. Surv.
Yathkyed lake	300	62 45	97 45	Geol. Surv.

Labrador et Ungava
Voir altitudes in Canada, p 226. J. White géogr.

MACKENZIE

Locality	Elev.	Lat.	Long.	Authority
Abbott lake..	1,320	63 50	106 20	Hanbury
Aberdeen lake.	130	64 30	100 00	Geol. Surv.
Artillery lake.	1,188	63 15	107 30	Interior
Aylmer, Lake	1,221	64 15	108 45	Interior
Bear rock	1,500	64 56	125 40	Geol. Surv.
Beverly lake...	133	64 40	101 00	Interior
Bois, Lac du	1,148	63 43	105 50	Interior
Campbell lake.	1,202	63 20	107 00	Interior
Clark mountain.	3,500	64 28	124 11	Geol. Surv.
Clinton-Colden lake.	1,221	64 00	107 30	Interior
Daly or Wholdaia lake.	1,200	60 30	104 00	Geol. Surv.
Doobaunt lake	500	63 00	101 30	Geol. Surv.
Douglas lake	1,343	63 15	107 25	Interior
Ennadai lake	1,100	60 50	101 30	Geol. Surv.
Great Bear lake.	391	66 00	120 00	Geol. Surv.
Great Slave lake.	520	62 00	114 00	Geol. Surv.
Kasba or White Partridge lake	1,270	60 30	102 25	Geol. Surv.
Rock-by-the-river-side	1,500	63 21	123 38	Geol. Surv.
Sandy lake.	940	63 53	104 55	Interior
Selwyn lake.	1,340	60 10	104 30	Geol. Surv.
Sifton lake	1,107	63 45	106 18	Interior
Wharton lake.	300	64 00	100 20	Geol. Surv.
White Partridge or Kasba lake	1,270	60 30	102 25	Geol. Surv.
Wholdaia or Daly lake.	1,200	60 30	104 00	Geol. Surv.

MANITOBA

Locality	Elev.	Lat.		Long.		Authority
		°	′	°	′	
Air Line junction..	761	49	55	97	11	C. P. R.
Alexander.	1,406	49	50	100	17	C. P. R.
Alpha	846	50	01	98	18	C. N. R.
Altamont		49	24	98	29	
Station	1,589					C. N. R.
Summit, 1·8 mile, W	1,626					C. N. R.
Altona	812	49	06	97	33	C. P. R.
Arden	1,087	50	16	99	16	C. P. R.
Argue	1,478	49	26	100	26	C. N. R.
Arnaud	796	49	15	97	06	C. P. R.
Assessippi	1,450	50	58	101	18	Geol. Surv.
Ashdown	1,343	49	32	99	37	C. N. R.
Ashville	1,129	51	10	100	18	C. N. R.
Austin	1,005	49	57	98	55	C. P. R.
Baden	1,035	52	47	101	13	C. N. R.
Bagot	935	49	58	98	37	C. P. R.
Baldur	1,408	49	23	99	13	C. N. R.
Balmoral	832	50	15	97	18	C. P. R.
Banting	1,242	49	38	99	40	C. P. R.
Barnsley	852	49	36	98	00	C. P. R.
Basswood	1,958	50	18	100	00	C. P. R.
Basswood lake	1,941	50	17	100	00	C. P. R.
Beauséjour	815	50	03	96	30	C. P. R.
Beaver	862	50	04	98	40	C. N. R.
Bedford	1,111	49	22	96	18	C. N. R.
Belmont	1,468	49	24	99	26	C. N. R.
Beresford	1,414	49	44	100	11	C. P. R.
Bergen	784	49	57	97	17	C. P. R.
Binscarth	1,725	50	37	101	16	C. P. R.
Birch River station	1,039	52	24	101	05	C. N. R.
Birds Hill station	761	49	50	97	00	C. P. R.
Birtle		50	26	101	02	
Station	1,712					C. P. R.
Birdtail brook, 3·1 miles, E., water, 1,554; rail	1,567					C. P. R.
Boissevain		49	14	100	02	
Station	1,683					C. P. R.
Summit, 1·1 mile, E., ground, 1,694; rail	1,690					C. P. R.
Bone lake	1,357	49	25	99	40	Upham
Bonnet lake	823	50	23	95	50	Dawson
Bowsman		52	14	101	13	
Station	1,016					C. N. R.
Bowsman river, water, 987; rail	1,003					C. N. R.
Brandon		49	51	99	57	
Can. Nor. station	1,255					C. N. R.
Can. Pac. station	1,194					C. P. R.
Assiniboine river at C. P. R. bridge, water	1,159					C. P. R.
Brandon hills	1,600	49	40	99	55	Upham
Buchan	845	50	06	96	18	C. P. R.
Burnside	872	49	58	98	26	C. P. R.
Carberry	1,258	49	52	99	20	C. P. R.
Carman	872	49	31	98	00	C. P. R.
Carroll	1,477	49	37	100	02	C. P. R.
Cartier	765	49	41	97	08	C. N. R.
Cartwright	1,531	49	06	99	19	C. P. R.
Chater	1,213	49	51	99	49	C. P. R.
Clearwater		49	07	99	02	
Station	1,498					C. P. R.
Clearwater brook, water, 1,426; rail	1,498					C. P. R.
Cowan	1,199	52	01	100	37	C. N. R.
Crandell	1,616	50	08	100	47	C. P. R.
Cross lake	1,046	49	50	95	14	C. P. R.
Cross Lake station	1,115	49	50	95	14	C. P. R.
Crystal City	1,513	49	09	98	56	C. P. R.
Culross	809	49	43	97	56	C. P. R.
Culver	1,005	49	54	95	40	C. P. R.
Curtis	827	49	56	98	05	C. N. R.
Cypress River station	1,232	49	33	99	05	C. P. R.

MANITOBA

Locality	Elev.	Lat.	Long.	Authority
		° ′	° ′	
Darlingford....		49 12	96 23	
Station......	1,561			C. P. R.
Summit, 3·5 miles, W.....	1,618			C. P. R.
Darwin......	972	49 55	95	C. P. R.
Dauphin.....	957	51 09	100	C. N. R.
Dauphin, Lake.....	869	51 20	99 46	Geol. Surv.
Deerwood	1,414	49 24	98 52	C. N. R.
Delean	1,419	49 35	100 33	C. P. R.
Deloraine	1,644	49 12	100 29	C. P. R.
Departure	1,433	49 27	99 23	C. N. R.
Dog lake....	815	51 00	98 30	C. P. R.
Dominion City		49 09	97 09	
Station......	787			C. P. R.
Roseau river, ice, 762 ; high water, 778 ; rail	787			C. P. R.
Douglas......	1,222	49 53	99 41	C. P. R.
Duck mountain......	2,600	51 45	100 50	Geol. Surv
Dufresne......	804	49 44	96 43	C. N. R.
Dufrost......	794	49 22	97 02	C. P. R.
Dunrea	1,498	49 24	99 43	C. N. R.
East Selkirk ..	714	50 08	96 49	C. P. R.
East Summit station.	1,766	50 15	99 45	C. P. R.
Edrans	1,068	50 03	99 08	C. P. R.
Elgin.. ...	1,530	49 26	100 15	C. N. R.
Elkhorn	1,630	49 58	101 14	C. P. R.
Elliott (now McCreary)	1,315	49 35	99 43	C. N. R.
Elm Creek station.....	820	49 41	98 00	C. P. R.
Elva	1,492	49 13	101 07	C. P. R;
Emerson..	792	49 00	97 12	C. P. R.
Emerson (West Lynne)	787	49 01	97 14	C. N. R.
Ethelbert	1,126	51 31	100 22	C. N. R.
Eustace	795	49 55	97 50	C. N. R.
Fairfax.. ..	1,569	49 26	100 07	C. N. R.
Fannystelle.. ..	789	49 45	97 47	C. P. R.
Findlay	1,429	49 34	100 44	C. P. R.
First Siding	1,544	49 16	98 47	C. P. R.
Fisher......		52 06	100 48	
Station.... .	1,130			C. N. R.
Sinclair river, 3·4 miles, S., bed, 1,131 ; rail......	1,138			C. N. R.
Fork River......		51 31	100 00	
Station	872			C. N. R.
Fork river, water, 851'; rail	869			C. N. R.
Forrest	1,428	49 58	99 55	C. P. R.
Fort Whyte	765	49 48	97 13	C. P. R.
Foxwarren......	1,751	50 30	101 08	
Station....	1,751			C. P. R.
Summit, 1·2 mile, E.......	1,752			C. P. R.
Franklin (formerly Bridge Creek)....	1,600	50 15	99 38	C. P. R.
Garland	1,127	51 39	100 26	C. N. R.
Gilbert Plains junction......	995	51 10	100 05	C. N. R.
Gilbert Plains station...	1,314	51 08	100 30	C. N. R.
Gladstone	884	50 14	98 56	C. P. R.
Glenboro	1,231	49 33	99 17	C. P. R.
Glencairn......	974	50 39	99 17	C. N. R.
Glenella......	954	50 33	99 10	C. N. R.
Glenlea	769	49 39	97 09	C. N. R.
Gonor	759	50 03	96 55	C. P. R.
Goodlands......	1,654	49 06	100 36	C. P. R.
Goose lake......	1,571	49 21	98 42	C. N. R.
Grandview	1,431	51 10	100 42	C. N. R.
Grass lake......	1,340	49 24	99 38	Upham
Greenway... .	1,401	49 22	99 06	C. N. R.
Gretna	829	49 00	97 33	C. P. R.
Griswold	1,417	49 46	100 27	C. P. R.
Gruber	847	51 37	99 56	C. N. R.
Indian Springs	1,471	49 23	98 53	C. N. R.
Johnston	1,660	50 38	101 19	C. P. R.
Julius......	928	50 00	96 12	C. P. R.
Kirkella......	1,684	50 02	101 22	C. P. R.

MANITOBA

Locality	Elev.	Lat.	Long.	Authority
		° ′	° ′	
Hamiota.	1,696	50 11	100 35	C. P. R.
Hargrave	1,579	49 55	101 02	C. P. R.
Harrowby	1,603	50 45	101 26	C. P. R.
Hartney...	1,420	49 29	100 31	C. P. R.
Haywood...	959	49 40	98 14	C. P. R.
Headingly.				
Can. Nor. station	778	49 50	97 21	C. N. R.
Can. Pac. Ry., South-western branch, crossing, 1·7 mile W...	779			C. N. R.
Can. Pac. station...	779	49 53	97 24	C. P. R.
High Bluff station...	829	50 01	98 09	C. P. R.
Hilton	1,370	49 30	99 31	C. N. R.
Holland	1,237	49 36	98 53	C. P. R.
Holmfield	1,551	49 08	99 28	C. P. R.
Horizon	840	50 00	98 23	C. N. R.
Kelloe...	1,823	50 29	100 46	C. P. R.
Kemnay	1,364	49 50	100 07	C. P. R.
Keyes (formerly Midway)... ...	975	50 14	99 06	C. P. R.
Kildonan	753	49 57	97 04	C. P. R.
Killarney	1,625	49 11	99 39	C. P. R.
Kranz hill... ...	2,200	49 02	99 59	Upham.
La Broquerie.	924	49 32	96 30	C. N. R.
Lac du Bonnet station..	852	50 16	96 03	C. P. R.
Larivière...		49 15	98 40	
Station..	1,326			C. P. R.
Pembina river, 1·3 mile W., water, 1,287; rail...	1,304			C. P. R.
La Salle.				
Can. Pac. station...	770	49 41	97 15	C. P. R.
La Salle river at Can. Pac. bridge, high water, 749; low water, 735; rail.	770			C. P. R.
Can. Nor. station...	789	49 54	97 46	C. N. R.
La Salle river at Can. Nor. bridge, bed, 779; rail ...	790			C. N. R.
Lauder	1,444	49 23	100 40	C. P. R.
Laurier	965	50 53	99 32	C. N. R.
Letellier... ...		49 09	97 18	
Station	784			C. N. R.
River aux Marais, high water, 776; low water, 763; rail.. ...	779			C. N. R.
Lizard lake...	1,577	49 17	98	C. P. R.
Lonely lake	815	51 09	98	C. P. R.
Long lake, high water, 803; low water. ...	798	50 06	97	C. P. R.
Long lake	1,929	50 18	100	C. N. R.
Lorette	790	49 47	96	C. N. R.
Lorne, Lake... ...	1,332	49 13	99	C. P. R.
Louise, Lake	1,332	49 14	99	C. P. R.
Lower Fort Garry.....	751	50 07	96	C. P. R.
Lowe...	791	49 22	97	C. N. R.
Macdonald..	838	50 03	98	C. P. R.
Mafeking... ...		52 41	101 2	
Station...	1,067			C. N. R.
Steep Rock river, bed, 1,024; rail... ...	1,037			C. N. R.
Makinak... ...	972	50 59	99 39	C. N. R.
Manitoba, Lake... ...		51 00	98 30	
High water. ...	814·5			C. P. R.
Low water.	808·5			C. P. R.
Mean water. ...	810			C. P. R.
Manitou...	1,589	49 14	98 32	C. P. R.
Marchand	1,001	49 26	96 22	C. N. R.
Margaret... ...	1,541	49 25	99 50	C. N. R.
Mariopolis	1,488	49 22	98 59	C. N. R.
Marquette.	807	50 04	97 42	C. P. R.
Martin	781	49 14	97 20	C. N. R.
Martinville.	1,216	49 46	99 49	C. N. R.
Mather...	1,527	49 06	99 11	C. P. R.
McCreary (formerly Elliott).. ...	992	50 45	99 27	C. N. R.
McGregor	955	49 58	98 45	C. P. R.
Meadow portage, summit	921	51 44	99 35	C. P. R.
Meadows	793	50 01	97 34	C. P. R.
Medora.	1,504	49 15	100 41	C. P. R.

MANITOBA

Locality	Elev.	Lat.		Long.		Authority
		° ′		° ′		
Melbourne...	1,248	49	52	99	11	C. P. R.
Melita......... ;		49	16	100	59	
Station......	1,410					C. P. R.
Souris river, high water, 1,405; low water, 1,387; rail	1,411					C. P. R.
Summit, 5·9 miles N	1,464					C. P. R.
Menteith station... .. ...:	1,305	49	33	100	26	C. P. R.
Menteith junction....	1,388	49	35	100	24	C. P. R.
Methven.........		49	37	99	42	
Station..	1,284					C. P. R.
Canadian Northern Ry. crossing, 1·6 mile W.. ...	1,311					C. P. R.
Miami..	1,012	49	22	98	14	C. N. R.
Millwood		50	40	101	24	
Station.......	1,376					C. P. R.
Assiniboine river, water, 1,321; rail. .... :.	1,348					C. P. R.
Milner........	913	50	10	96	15	C. P. R.
Miniota	1,500	50	08	101	00	C. P. R.
Minitonas....	1,086	52	05	101	00	C. N. R.
Minnedosa		50	15	99	50	
Station..............	1,671					C. P. R.
Little Saskatchewan river, 0·1 mile E., water, 1,658, rail.	1,670					C. P. R.
Little Saskatchewan river, 1·7 mile S., water, 1,645, rail	1,659					C. P. R.
Minto. :....		49	25	100	00	
Station..	1,593					C. N. R.
Summit, 1·0 mile, W......	1,597					C. N. R.
Molson..	879	50	02	96	18	C. P. R.
Morden.. . ...|	988	49	12	98	05	C. P. R.
Morris..		49	21	97	22	
Can. Pac. station............	773					C. P. R.
Can. Nor. station......... ...	773					C. N. R.
Morris river, high water, 771; low water	744					C. P. R.
Murray Park..	780	49	54	97	17	C. P. R.
Myrtle	827	49	22	97	50	C. N. R.
Napinka..................	1,466	49	19	100	49	C. P. R.
Naughton (Rounthwaite).....	1,307	49	40	98	47	C. N. R.
Neepawa.	1,209	50	14	99	26	C. P. R.
Nesbitt	1,407	49	36	99	52	C. P. R.
Newdale		50	21	100	11	
Station..	1,984					C. P. R.
Summit, 4·0 miles E., ground, 1,999; rail...... ..	1,992					C. P. R.
Ninette... :		49	24	99	38	
Station	1,363					C. N. R.
Depression, 0·5 mile, E...............	1,353					C. N. R.
Ninga........	1,649	49	14	99	52	C. P. R.
Niverville............	776	49	36	97	02	C. P. R.
Novra............. ...:	1,059	52	32	101	04	C. N. R.
Oak Lake station......	1,415	49	46	100	37	C. P. R.
Oakland..	819	50	07	98	18	C. N. R.
Oak River........		50	08	100	26	
Station..	1,711					C. P. R.
Oak river, bed, 1,675; rail:....	1,712					C. P. R.
Oakville.........	815	49	56	98	00	C. N. R.
Ochre River......		51	03	99	46	
Station.....	929					C. N. R.
Ochre river, high water, 914; rail...	917					C. N. R.
Ogilvie.	925	50	17	99	03	C. N. R.
Ontop. ...	1,493	49	25	99	40	C. N. R.
Osborne......	777	49	33	97	21	C. P. R.
Otterburne....		49	31	97	02	
Station......	780					C. P. R.
Rat river, ig water, 764; low water, 753; rail.....	779					C. P. R.
Parkdale.....h . h.....	755	50	02	97	02	C. P. R.
Pelican lake.	1,340	49	20	99	30	C. N. R.
Pembina mountain, on Inter. boundary, crest of escarpment........	1,695	49	00	98	10	Bdy. Com.

MANITOBA

Locality	Elev.	Lat.		Long.		Authority
		°	′	°	′	
Pettapiece ..	1,695	50	06	100	11	C. P. R.
Pierson	1,531	49	11	101	15	C. P. R.
Pilot Mound	1,551	49	12	98	53	C. P. R.
Pine Creek station	985	50	02	99	00	C. P. R.
Pine River		51	48	100	31	
Station	1,146					C. N. R.
Pine river. bed, 1,143; rail	1,154					C. N. R.
Pipestone	1,451	49	33	100	57	C. P. R.
Plumas	927	50	23	99	04	C. N. R.
Plum Coulée station	834	49	12	97	45	C. P. R.
Poplar Point	815	50	04	97	58	C. P. R.
Porcupine mountain	2,500	52	30	101	45	Geol. Surv.
Portage junction	760	49	51	97	09	C. N. R.
Portage la Prairie		49	59	98	17	
Can. Pac. station	854					C. P. R.
Can. Nor. station	854					C. N. R.
Assiniboine river, 4·8 miles E., high water, 842; rail	847					C. N. R.
Powell	1,035	52	50	101	24	C. N. R.
Purves	1,574	49	07	98	43	C. P. R.
Raith	970	49	54	95	42	C. P. R.
Rapid City junction	1,557	50	07	100	04	C. P. R.
Rapid City		50	06	100	02	
Station	1,580					C. P. R.
Little Saskatchewan river, water, 1,570; rail	1,580					C. P. R.
Rathwell		49	39	98	32	
Station	1,070					C. P. R.
Morris river, 1·6 mile E., low water, 1,034; rail	1,047					C. P. R.
Raven lake	1,791	50	22	100	36	C. P. R.
Reaburn	806	50	05	97	51	C. P. R.
Rennie	1,052	49	51	95	33	C. P. R.
Reston	1,513	49	33	101	05	C. P. R.
Ridgeway	841	50	01	98	30	C. N. R.
Riding mountain	2,000	51	00	100	30	Geol. Surv.
Riverdale	1,637	50	12	99	57	C. P. R.
Rock lake, high water, 1,330; ice	1,321	49	13	99	10	C. P. R.
Roland	855	49	22	97	56	C. N. R.
Rosebank	910	49	22	98	06	C. N. R.
Roseau swamp	1,004	49	00	96	08	Bdy. Com.
Rosenfeld	796	49	13	97	33	C. P. R.
Rosser	796	49	59	97	26	C. P. R.
Rounthwaite (Naughton)	1,307	49	40	98	47	C. N. R.
Routledge	1,420	49	47	100	47	C. P. R.
Russell	1,832	50	46	101	17	C. P. R.
Ste. Agathe	774	49	34	97	13	C. N. R.
Ste. Anne		49	40	96	39	
Station	827					C. N. R.
Seine river, 0·9 mile W., high water, 819; low water, 810; rail	824					C. N. R.
St. Boniface	756	49	53	97	06	C. P. R.
St. Claude	984	49	39	98	19	C. P. R.
St. James		49	53	97	10	
Station	764					C. P. R.
Assiniboine river, high water, 743; low water, 734; rail	764					C. P. R.
St. Jean	775	49	16		20	C. P. R.
St. John, Butte or Kranz hill	2,200	49	02		59	Upham
St. Martin, Lake	795	51	40		30	Geol. Surv.
St. Norbert	762	49	46		09	C. N. R.
St. Norbert	769	49	46		02	C. P. R.
Sandilands	1,161	49	20	96	18	C. N. R.
Sclater	1,149	51	56	100	36	C. N. R.
Sewell	1,255	49	53	99	31	C. P. R.
Shanawan	774	49	37	97	19	C. P. R.
Shelley	929	49	59	96	06	C. P. R.
Shellmouth		50	57	101	27	
Bottom of valley	1,335					Geol. Surv.
Top of bank	1,565					Geol. Surv.
Shoal Lake station	1,821	50	26	100	35	C. P. R.

MANITOBA

Locality	Elev.	Lat.		Long.		Authority
		°	′	°	′	
Shoal lake.	1,793	50	24	100	35	C. P. R.
Shoal lake, high water.	856	50	25	97	40	C. P. R.
Sidney	1,232	49	54	99	05	C. P. R.
Sifton junction.	959	51	23	100	08	C. N. R.
Sifton station.	967	51	21	100	08	C. N. R.
Silver Plains.	776	49	28	97	17	C. N. R.
Sinclair.	1,656	49	34	101	17	C. P. R.
Sinnot		50	03	96	27	
Siding	798					P. R.
Brokenhead river, water, 784; rail.	796					P. R.
Six-mile	843	50		98	22	P. R.
Snowflake.	1,563	49		98	38	P. R.
Solsgirth	1,798	50		100	55	C. P. R.
Somerset	1,565	49	01	98	30	C. N. R.
Souris.		49	28	100	15	
Station	1,400					P. R.
Souris river, water.	1,164					P. R.
Southwestern Branch junction	760	49	54	97	09	P. R.
Sprague	1,063	49	02	95	39	C. N. R.
Starbuck.	781	49	47	97	36	C. P. R.
Steinbach	881	49	34	96	33	C. N. R.
Stockton	1,190	49	35	99	26	C. P. R.
Stonewall.	826	50	08	97	19	C. P. R.
Stony Mountain station	771	50	04	97	12	C. P. R.
Strathclair.	1,910	50	24	100	23	C. P. R.
Summit.		49	12	95	58	
Station	1,237					C. N. R.
Summit, 0·8 mile, W.	1,245					C. N. R.
Swan Lake station.	1,557	49	25	98	47	C. N. R.
Swan lake.	1,310	49	20	98	55	Upham
Swan lake.	817	50	50	98	10	C. P. R.
Swan lake	855	52	30	100	45	Geol. Surv.
Swan River station	1,112	52	06	101	16	C. N. R.
Telford	1,108	49	50	95	19	C. P. R.
Teulon (formerly Foxton)	862	50	23	97	15	C. P. R.
Thornhill	1,314	49	12	98	13	C. P. R.
Tiger hills.	1,500-1,600	49	30	99	00	Upham
Tiger hill, Big.	1,640	49	30	99	45	Upham
Tiger Hills		49	26	99	34	
Station	1,521					C. N. R.
Summit, 0·2 mile, W	1,527					C. N. R.
Townline	827	50	04	98	18	C. N. R.
Trackend (now Argue)	1,478	49	26	100	26	C. N. R.
Treesbank		49	38	99	36	
Station	1,206					C. P. R.
Souris river, low water, 1,113; rail.	1,169					C. P. R.
Treherne		49	37	98	41	
Station	1,211					C. P. R.
Morris river, low water, 1,166; rail	1,222					C. P. R.
Turtle mountain	2,300	49		100	43	Geol. Surv.
Tyndall.	795	50		96	39	C. P. R.
Underhill.	1,495	49	04	100	23	C. N. R.
Union Point.	776	49	56	97	15	C. N. R.
Valley River.		51	15	100	08	
Station	971					C. N. R.
Valley river, high water, 956; rail.	970					C. N. R.
Varcoe.		50	04	99	57	
Station	1,725					C. P. R.
Summit, 2·9 miles, N.	1,805					C. P. R.
Vassar.	1,168	49	06	95	50	C. N. R.
Victoria Park	757	50	04	96	59	C. P. R.
Virden.	1,444	49	51	100	54	C. P. R.
Waskada.	1,551	49	06	100	50	C. P. R.
Waterhen lake.	822	52	00	99	35	Geol. Surv.
Wawanesa	1,198	49	35	99	41	C. N. R.
Wellwood	1,282	50	02	99	19	C. P. R.
Westbourne		50	08	98	33	
Station	832					C. P. R.
Whitemud river, bed, 812; rail.	832					C. P. R.

MANITOBA

Locality	Elev.	Lat.		Long.		Authority
		′	″	′		
Westgate.........	1,011	52	50	101	32	C. N. R.
West Lynne		49	01	97	14	
Can. Pac., old station......	791					C. P. R.
Can. Nor. station (now Emerson)......	787					C. N. R.
West Selkirk....		50	09	96	53	
Station	733					C. P. R.
General prairie level............	737					C. P. R.
Red river, extreme high water (1826), 731 ; (1852),						
725 ; ice (1876)	711					C. P. R.
Whitemouth................................		49	58	96	00	
Station	908					C. P. R.
Whitemouth river, ice, 877 ; flood, 896 ; high water,						
888 ; rail	900					C. P. R.
White Plains	801	49	51	97	32	C. N. R.
Whitewater.....................................	1,659	49	12	100	16	C. P. R.
Whitewater lake, high water, 1,637 ; low water..	1,632	49	15	100	20	C. P. R.
Winkler....	988	49	11	97	56	C. P. R.
Winnipeg		49	54	97	07	
Can. Pac. station	757					C. P. R.
Can. Nor. station	753					C. N. R.
Main st. bridge, top of pier, S. end	751					City Eng.
Catch basin, opposite 318 Main st.	757					City Eng.
Receiving reservoir, bottom	751					City Eng.
City hall, basement floor.........	755					City Eng.
City hall, main floor	768					City Eng.
Portage ave., at city limits...................	762					City Eng.
Red river, at mouth of Assiniboine river, extreme						
high water (1826), 762 ; (1852), 760 : (1860), 758 ;						
(1875), 743 ; ice (1876).	725					C. P. R.
Winnipeg junction..............................	757	49	54	97	06	C. P. R.
Winnipeg, Lake		52	00	98	00	
High water..........	713					C. P. R.
Low water....	708					C. P. R.
Mean water.................................. ...	710					C. P. R.
Winnipegosis........................ ...	839	51	39	99	55	C. N. R.
Winnipegosis, Lake..	828	52	30	100	00	C. P. R.
Wood Bay station..	1,545	49	15	98	48	C. P. R.
Woodridge............. \	1,223	49	17	96	10	C. N. R.
Woodside		50	11	98	46	
Station	859					C. P. R.
Whitemud river, bed, 849 ; rail...............	860					C. P. R.
Youill...........	854	50	02	98	36	C. N. R.

NEW BRUNSWICK

Locality	Elev.	County	Authority
Aboushagon	216	Westmorland	Geol. Surv.
Acadieville		Kent	
Siding	306		I.C.R.
Summit, 0·5 mile N	312		I.C.R.
Acamac	45	St. John	C.P.R.
Adamsville	300	Kent	I.C.R.
Adder lake	1,350	Northumberland	Ganong
Allandale settlement	575	York	Geol. Surv.
Alma	24	Albert	A.S.R.
Anagance	159	Kings	I.C.R.
Andover	257	Victoria	C.P.R.
Annidale		Queens	
Station	524		C.R. of N.B.
Summit, 1·4 mile S., ground, 534 ; rail	544		C.R. of N.B.
Apohaqui	73	Kings	I.C.R.
Arbuckle		Victoria	
Station	370		C.P.R.
Wapskehegan river, extreme freshet, 357 ; ordinary water, 353 ; rail	365		C.P.R.
Aroostook	271	Victoria	C.P.R.
Aroostook Portage P. O.	625	Victoria	Geol. Surv.
Arthurette	343	Victoria	C.P.R.
Ashland	510	Carleton	Geol. Surv.
Astle		Northumberland	
Station	479		C.E.R.
Summit, 1·1 mile, N	532		C.E.R.
Aulac	26	Westmorland	I.C.R.
Baie Verte	10	Westmorland	N.B & P.E.I.R.
Bailey	29	Sunbury	C.P.R.
Bailey, Mount	1,714	Restigouche	Ganong
Bald head or Riley mountain	1,866	Victoria	Geol. Surv.
Bald hill	700	Queens	Geol. Surv.
Bald or Wadawamketch mountain	900	York	Geol. Surv.
Bald mountain (South branch Nipisiguit river)	2,500	Northumberland	Geol. Surv.
Bald mountain (Nipisiguit river)	1,922	Northumberland	Geol. Surv.
Bald or Sagamook mountain	2,604	Restigouche	Bdy. Com.
Bald mountain	1,120		Geol. Surv.
	1,430	Kings	Ganong
	1,390		Murdoch
Ballentine (now Woolastook)	33	Kings	C.P.R.
Barber Dam		Charlotte	
Station	411		C.P.R.
Digdeguash river, 1·4 mile N., water, 394 ; rail	400		C.P.R.
Barnaby River		Northumberland	
Station	56		I.C.R.
Summit, 2·5 miles, N	141		I.C.R.
Barnesville	470	Kings	Geol. Surv.
Bar Road station	29	Charlotte	C.P.R.
Bartibog		Gloucester	
Station, (summit)	521		I.C.R.
Bartibog river, 1·0 mile N., bed, 513 ; rail	518		I.C.R.
Bartibog settlement	250	Northumberland	Geol. Surv.
Bartlett	113	Charlotte	C.P.R.
Bass River settlement	155	Kent	Geol. Surv.
Bass River static	50	Gloucester	Car. R.
Bath	201	Carleton	C.P.R.
Bathurst		Gloucester	
Intercolonial station	45		I.C.R.
Car. station	69		Car. R.
Beaver Brook station	331	Northumberland	I.C.R.
Beech hill	400	Westmorland	Geol. Surv.
Beechwood	218	Carleton	C.P.R.
Belledune	84	Gloucester	I.C.R.
Belleisle		Kings	
Station	216		C.R. of N.B.
Belleisle creek, 0·8 mile S., high water, 142 ; bed, 133 ; rail	165		C.R. of N.B.
Benton	406	York	C.P.R.
Berrys Mills	215	Westmorland	I.C.R.

NEW BRUNSWICK

Locality	Elev.	County	Authority
Birch ridge	840	Victoria	Col. Surv.
Bishopland	286	Kent	Col. Surv.
Black River settlement	100	Northumberland	Geol. Surv.
Blackville	66	Northumberland	C.E.R.
Blagdon	50	Kings	C.P.R.
Blair, Mount	610	Charlotte	Admir. chart
Blissfield	226	Northumberland	C.E.R.
Bloomfield	36	Kings	I.C.R.
Bloomfield corner	515	Carleton	Geol. Surv.
Blue mountain	1,367	Victoria	Bdy. Com.
Boiestown	210	Northumberland	C.E.R.
Bonny River station	72	Charlotte	S.L.R.
Boundary Creek station	79	Westmorland	I.C.R.
Boyne settlement	365	Sunbury	Geol. Surv.
Bridgetown	49	Gloucester	Car. R.
Briggs corner	540	Carleton	Geol. Surv.
Bristol	183	Carleton	C.P.R.
Brooklyn	85	Westmorland	N.B. & P.E.I.R.
Brookville	28	St. John	I.C.R.
Browne	261	Carleton	C.P.R.
Buctouche		Kent	
Station	17		B. & M. R.
Wharf	12		B. & M. R.
Bull Moose hill	750	Kings	Geol. Surv.
Burnside	216	York	C.P.R.
Burnsville		York	
Station	46		Car. R.
Caraquet river, bed, 18; rail	46		Car. R.
Burtt Lake		York	
Station	520		C.P.R.
Summit, 0·2 mile, W	538		C.P.R.
Cahill	466	York	C.P.R.
Caledonia mountain	1,240	Albert	Geol. Surv.
Calhoun	54	Westmorland	I.C.R.
California P. O.	670	Victoria	Geol. Surv.
Callina corner	475	Kings	Geol. Surv.
Campbellton	40	Restigouche	I.C.R.
Campbellton settlement	660	York	Geol. Surv.
Canaan	263	Westmorland	I.C.R.
Canaan settlement	540	Carleton	Geol. Surv.
Canterbury		York	
Station	556		C.P.R.
Summit, 0·5 mile, N	565		C.P.R.
Cape Bald settlement	100	Westmorland	Geol. Surv.
Cape Breton	129	Westmorland	B. & M. R.
Cape Tormentine wharf	13	Westmorland	N.B. & P.E.I.R.
Caraquet	81	Gloucester	Car. R.
Cardigan	60	York	C.P.R.
Carleton	16·5	St. John	I.C.R.
Carleton, Mount	2,716	Northumberland	Bdy. Com.
Carlow, cross roads	755	Carleton	Geol. Surv.
Carroll	200	Northumberland	C.E.R.
Carroll ridge	1,009	York	Geol. Surv.
Carrs Mill	578	Carleton	C.P.R.
Case Settlement station	272	Kings	C.R. of N.B.
Centreville	375	Carleton	Geol. Surv.
Chamcook	86	Charlotte	C.P.R.
Chamcook hill	637	Charlotte	Admir. chart
Charlo	53	Restigouche	I.C.R.
Chatham		Northumberland	
Station	99		C.E.R.
Wharf	7		C.E.R.
Chatham junction		Northumberland	
Station	130		I.C.R.
Southwest branch Miramichi river, ordinary high water, 2·5; rail	33		I.C.R.
Cherry hill	808	York	Geol. Surv.
Chipman	36	Queens	C.R. of N.B.
Chiputneticook lake	378	York	Coast Surv.

NEW BRUNSWICK

Locality	Elev.	County	Authority
Church hill	1,200	Albert	Geol. Surv.
Claire	540	Madawaska	Tem. R.
Clarendon		Queens	
Station	185		C.P.R.
Summit, 0·2 mile E., ground, 200; rail	194		C.P.R.
Clifton	119	Gloucester	Car. R.
Clearwater		York	
Station	485		C.E.R.
Summit, 3·1 miles S	566		C.E.R.
Clones settlement	400	Queens	Geol. Surv.
Cloverdale	350	Carleton	Geol. Surv.
Coal Branch	218	Kent	I.C.R.
Coal Creek station	61	Queens	C.R. of N.B.
Cocagne	62	Kent	B. & M.R.
Coldbrook	15	St. John	I.C.R.
College Bridge	32	Westmorland	I.C.R.
Connors	509	Madawaska	Tem. R.
Cork		York	
Station	420		C.P.R.
Summit, 0·9 mile E.	447		C.P.R.
Cottrell	476	York	C.P.R.
County Line station	529	York	C.P.R.
Covered Bridge station	159	York	C.E.R.
Cranberry lake	436	York	Geol. Surv.
Cripp hill	1,225	Kings	Geol. Surv.
Cross Creek settlement	715	York	Geol. Surv.
Cross Creek station	206	York	C.E.R.
Cumberland	64	Queens	C.R. of N.B.
Currie	282	Victoria	C.P.R.
Dalhousie junction		Gloucester	
Station	79		I.C.R.
Summit siding, 2·0 miles E., rail	230		I.C.R.
Dalhousie station	13	Gloucester	I.C.R.
Dalhousie, Mount	715	Restigouche	Admir. chart
Darby Gillans	400	Sunbury	Geol. Surv.
Debec		Carleton	
Station	539		C.P.R.
Summit, 1·9 mile S	599		C.P.R.
Deep Creek station	131	Carleton	C.P.R.
Deer Lake station	531	York	C.P.R.
De Merchant	323	Victoria	C.P.R.
Denys, Mount	1,922	Northumberland	Geol. Surv.
Derby	33	Northumberland	I.C.R.
Dibblee	303	Carleton	C.P.R.
Doak	125	York	C.P.R.
Doaktown		Northumberland	
Station	101		C.E.R.
Southwest Miramichi river, 0·8 mile E., water 72; rail	89		C.E.R.
Doherty corner	322	Sunbury	Geol. Surv.
Dorchester	27	Westmorland	I.C.R.
Dorchester, Cape	300	Westmorland	Geol. Surv.
Dorchester Road	107	Westmorland	I.C.R.
Dougherty	141	Charlotte	C.P.R.
Douglas		York	
Station	52		C.P.R.
Nashwaaksis river, water 28; rail	41		C.P.R.
Douglas lake	290	St. John	Geol. Surv.
Dumbarton	224	Charlotte	C.P.R.
Dunsinane		Kings	
Siding	157		I.C.R.
Summit, 0·5 mile E	167		I.C.R.
Durham	84	York	C.E.R.
Eagle mountain	854	Charlotte	Ganong
Eagle rock	75	Queens	C.P.R.
East Scotch settlement	520	Kings	Geol. Surv.
Edmundston	468	Madawaska	C.P.R.
Eel River station	28	Restigouche	I.C.R.
Eel River lake, First	520	York	Geol. Surv.

NEW BRUNSWICK

Locality	Elev.	County	Authority
Eel River lake, Second......	550	York	Geol. Surv.
Elgin corner	373	Albert	Geol. Surv.
Emigrant settlement.............	135	Westmorland	Geol. Surv.
Enaud, Mount or Baldface mountain.................	1,600	Northumberland	Ganong
Enniskillen	113	Queens	C. P. R.
Evans....	223	Westmorlan'	I.C. R.
Fairville	103	St. John	C. P. R.
Fall Brook station.	561	Carleton	C. P. R.
Farmerston	305	Carleton	Geol. Surv.
Fawcett Mill....................	146	Westmorland	E. & H. R.
Fenwick..	605	Kings	Geol. Surv.
Flagstaff hill.............	585	Carleton	Geol. Surv.
Flatlands.	63	Restigouche	I. C. R.
Florenceville	175	Carleton	C. P. R.
Folly point.... ...	315	Westmorland	Geol. Surv.
Found head	380	St. John	Admir. chart
Four Falls	355	Victoria	Geol. Surv.
Franquelin hill........	1,239	Restigouche	Ganong
Fredericton		York	
Station .	41		C. P. R.
St. John river, high water, 18; ordinary freshet level, 29; extreme high water (1887), 35; extreme low water, 10; rail, centre of draw span	41		C. P. R.
Almshouse, floor level	77		City Eng.
B.M., top of stone post N.W. corner Normal school lot	40·6		City Eng.
B.M., on base of lamp post, Smith st. and Woodstock road....	39·5		City Eng.
Top of rubble foundation wall, waterworks building	37·6		City Eng.
Average level of city....	38		City Eng.
Bottom of waterworks pump well..........	2		City Eng.
Bed of river, side of crib work.	3		City Eng.
Fredericton junction.....		Sunbury	
Station ...	71		C. P. R.
N.W. branch Oromocto river, 0·2 mile E., high water 40; low water 32; rail	65		C. P. R.
Furbish pond	1,470	Victoria	Ganong
Galloway settlement.	110	Kent	Geol. Surv.
Gardens hill	1,164	Restigouche	Ganong
Gaspereaux..	147	Queens	C. P. R.
George lake...........	430	York	Geol. Surv.
Gibson		York	
Can. Pac. station.........	35		C. P. R.
Can. East. station ...	35		C. E. R.
Gilmor Mills......	510	Carleton	Geol. Surv.
Girvan settlement..	138	Kent	Geol. Surv.
Glasier	73	York	C. P. R.
Glassville.......	770	Carleton	Geol. Surv.
Gloucester junction	109	Gloucester	I. C. R.
Golden mountain	1,150	Albert	Geol. Surv.
Golden ridge..	1,500	Carleton	Geol. Surv.
Gordon, Mount... ...	1,552	Restigouche	Ganong
Gordonvale settlement..	830	York	Geol. Surv.
Gordonville....	840	Carleton	Geol. Surv.
Grand Anse....	79	Gloucester	Car. R.
Grand Bay... .	65	Kings	C. P. R.
Grand Falls..		Victoria	
Station... .	498		C. P. R.
Summit, 0·3 mile S...	506		C. P. R.
St. John river, 0·9 mile N., low water 413; rail....	464		C. P. R.
Grand lake	427	York	Coast Surv.
Grand River station.........	432	Madawaska	C. P. R.
Grant	545	York	C. P. R.
Grays lake	1,370	Victoria	Ganong
Greenfield....	775	Carleton	Geol. Surv.
Greenfield settlement	175	Sunbury	Geol. Surv.
Green Hill settlement....	725	York	Geol. Surv.
Green Point...........	327	York	C. P. R.
Green River lake..............	1,362	Restigouche	I. C. R.

NEW BRUNSWICK

Locality	Elev.	County	Authority
Green River mountain	1,600	Madawaska	Bdy. Com.
Green River		Madawaska	
Station	478		C. P. R.
Green river, 0·8 mile E., high water 448 ; rail	451		C. P. R.
Green Road (formerly Greenville)	582	Carleton	C. P. R.
Griffin	284	Victoria	C. P. R.
Gull lake	795	Victoria	I. C. R.
Gulquac lake	1,330	Victoria	Ganong
Hainesville	384	York	C. P. R.
Hale	189	Carleton	C. P. R.
Hall hill	150	Westmorland	Geol. Surv.
Hampton	27	Kings	I. C. R.
Harcourt		Kent	
Station	200		I. C. R.
Richibucto river, 1·8 mile N., water 97 ; rail	129		I. C. R.
Hardingville	430	St. John	Geol. Surv.
Hardwood Ridge settlement	400	Sunbury	Geol. Surv.
Hardy Corner	57	Westmorland	N. B. & P. E. I. R.
Hartland	151	Carleton	C. P. R.
Harvey		York	
Station	493		C. P. R.
Summit, 1·8 mile W	519		C. P. R.
Havelock	291	Kings	E. & H. R.
Havelock (now Shewen)		Carleton	
Station	631		C. P. R.
Summit, 0·2 mile E	641		C. P. R.
Hawkshaw	8'	York	St. J. V. & R. I. R.
Headline settlement	49	Queens	Geol. Surv.
Head, Mount	2,439	Northumberland	Ganong
Henderson corner	630	Carleton	Geol. Surv.
Hewitt	205	Charlotte	C. P. R.
Hibernia	700	Queens	Geol. Surv.
Hillside	308	Victoria	C. P. R.
Hind lake	1,450	Victoria	Ganong
Honeydale	254	Charlotte	C. P. R.
Hood	426	York	C. P. R.
Howland ridge	945	York	Geol. Surv.
Hoyt	57	Sunbury	C. P. R.
Humphrey	56	Westmorland	I. C. R.
Inches	269	York	C. P. R.
Indian lake	1,405	Victoria	Ganong
Indian or Lute mountain	450	Westmorland	Geol. Surv.
Indiantown	30	Northumberland	I. C. R.
Indian village	36	York	St. J. V. & R. I. R.
Ingleside	75	Kings	C. P. R.
Inkerman	15	Gloucester	G. S. R.
Intervale		Westmorland	
Station	124		E. & H. R.
North river, water 101 ; rail	122		E. & H. R.
Irishtown		Westmorland	
Station	237		B. & M. R.
Summit, 0·5 mile S.	255		B. & M. R.
Village	400		Geol. Surv.
Iron-ore hill	630	Carleton	Geol. Surv.
Ivy corner	670	Carleton	Geol. Surv.
Jacksonville	360	Carleton	Geol. Surv.
Jacquet River station	55	Restigouche	I. C. R.
Janeville	44	Gloucester	Car. R.
Jeffrey	375	Kings	Geol. Surv.
Johnville	640	Carleton	Geol. Surv.
Jordan mountain	775	Kings	Geol. Surv.
Jubilee	47	Kings	I. C. R.
Kars	288	Kings	Geol. Surv.
Kenney Mill	163	Albert	A. S. R.
Kent junction	267	Kent	I. C. R.
Keswick	38	York	C. P. R.
Ketepec	75	St. John	C. P. R.
Kilburn	271	Victoria	C. P. R.
Killam Mill	266	Westmorland	E. & H. R.

NEW BRUNSWICK

Locality	Elev.	County	Authority
Kingston peninsula............		Kings	Geol. Surv.
Knowlesville..............		Carleton	Geol. Surv.
Lakefield..............................	500	Kings	Geol. Surv.
Lakeside		Kent	I. C. R.
Lake settlement............................		Kent	Geol. Surv.
Lanes Quarry.............................	90	Westmorland	N. B. & P. E. I. R.
Latimore lake..............................	102	St. John	Geol. Surv.
La Tour, Mount...............		Northumberland	Ganong
Lawrence..........		Charlotte	C. P. R.
L'Eglise		Madawaska	C. P. R.
Lepreau		Charlotte	S. L. R.
Limestone...........................			C. P. R.
Lindsay...	450	Carleton	Geol. Surv.
Langley.................	30		C. P. R.
Little Bald mountain (N.W. Miramichi).............	2,0..	Northumberland	Geol. Surv.
Little River station............................	6.	Kent	B. & M. R.
Loch Lomond.....................................	302	St. John	Geol. Surv.
Loggieville	53	Northumberland	C. E. R.
Long lake......................................	1,370	Victoria	Geol. Surv.
Long settlement..................................	1,200	Kings	Geol. Surv.
Lower Caverhill settlement.................	700	York	Geol. Surv.
Lower French village.............................	42	York	St. J. V. & R. L. R.
Lower Prince William...................	63	York	St. J. V. & R. L. R.
Ludlow..................................	151	Northumberland	C. E. R.
Lute or Indian mountain	450	Westmorland	Geol. Surv.
McAdam.........................		York	
Station	445		C. P. R.
Summit, 0.8 mile W................	522		C. P. R.
McCafferty settlement.........	455	Carleton	Geol. Surv.
McDougall	195	Westmorland	B. & M. R.
McInnes lake	1,465	Northumberland	Ganong
McLeod hill...............	416	York	Geol. Surv.
McNamee	133	Northumberland	C. E. R.
Magaguadavic........................		York	
Station	391		C. P. R.
Magaguadavic river, 1.7 mile E., high water; 372; low water 369; rail........	409		C. P. R.
Magaguadavic lake	377	York	Geol. Surv.
Manners Sutton.........................	575	York	Geol. Surv.
Manzer......................	64	York	C. E. R.
Maple ridge........................	875	York	Geol. Surv.
Mapleton............................	600	York	Geol. Surv.
Maringouin, Cape..........	220	Westmorland	Admir. chart
Markhamville, manganese mine	700	Kings	Geol. Surv.
Marks hill	290	Charlotte	Admir. chart
Martin................................	460	Madawaska	C. P. R.
Marysville...........................	60	York	C. E. R.
Matthews head.......................	712	Albert	Geol. Surv.
Maudslay	417	York	C. P. R.
Maxwell..............................	181	Charlotte	C. P. R.
Meadow Brook station...............	92	Westmorland	I. C. R.
Meadows (now Honeydale)	254	Charlotte	C. P. R.
Mechanic Settlement, hill 4 miles S. W.	1,400	Kings	Geol. Surv.
Melrose.............................	101	Westmorland	N. B. & P. E. I. R.
Memramcook	28	Westmorland	I. C. R.
Menzie settlement...................	150	St. John	Geol. Surv.
Merithew lake.......................	1,390	Victoria	Ganong
Middle Sackville	26	Westmorland	N. B. & P. E. I. R.
Midgic................................	75	Westmorland	N. B. & P. E. I. R.
Millerton............................	23	Northumberland	I. C. R.
Millicete............................	268	Victoria	C. P. R.
Milltown junction....................	51	Charlotte	C. P. R.
Milltown station.....................	51	Charlotte	C. P. R.
Millville.............................	485	York	C. P. R.
Milnagek or Island lake.............	1,510	Victoria	Ganong
Milpugos or Mud lake.............	1,533	Victoria	Ganong
Miramichi lake......................	750	York	I. C. R.
Model Farm	130	Kings	I. C. R.
Moffatt..............................	33	Restigouche	I. C. R.

NEW BRUNSWICK

Locality	Elev.	County	Authority
Moncton		Westmorland	
Station	50		I. C. R.
Wharf	29		B. & M. R.
Moore Mills	291	Charlotte	C. P. R.
Moose mountain	1,490	Carleton	Geol. Surv.
Morrison	76	York	C. P. R.
Mount Hope	400	York	Geol. Surv.
Mount Uniacke	210	Westmorland	Geol. Surv.
Mud lake	418	York	Coast Surv.
Muniac	229	Victoria	C. P. R.
Murchie	387	York	C. P. R.
Nackawic	464	York	C. P. R.
Nalaisk mountain	2,520	Northumberland	Ganong
Nashe Creek station	25	Restigouche	I. C. R.
Nashwaak	67	York	C. E. R.
Nason	38	Sunbury	C. P. R.
Nauwigewauk	23	Kings	I. C. R.
Nelson station	127	Northumberland	C. E. R.
Nelson junction	84	Northumberland	C. E. R.
Nerepis	30	Kings	C. P. R.
New Bandon	117	Gloucester	Car. R.
Newburg	206	Carleton	C. P. R.
Newburg settlement	565	Carleton	Geol. Surv.
Newcastle	64	Queens	C. R. of N. B.
Newcastle		Northumberland	
Station	134		I. C. R.
Northwest branch Miramichi river, ordinary high			I. C. R.
water 2·5; extreme high water 5; rail	29		
New Horton	270	Albert	Geol. Surv.
New Ireland	1,100	Albert	Geol. Surv.
New Mills	37	Restigouche	I. C. R.
New Zealand settlement	465	York	Geol. Surv.
Nictau	576	Victoria	Bdy. Com.
Nictor lake	886	Restigouche	Bdy. Com.
Nipisiguit lake, First	996	Northumberland	Geol. Surv.
Nipisiguit lake, Second	980	Northumberland	Geol. Surv.
Nixon	151	Carleton	C. P. R.
North lake	427	York	Coast Surv.
Norton	53	Kings	I. C. R.
Nortondale settlement	470	York	Geol. Surv.
Notre Dame	80	Kent	B. & M. R.
Oak Bay station	100	Charlotte	S. L. R.
Odell River station	345	Victoria	C. P. R.
Olinville	400	Queens	Geol. Surv.
Ononette	50	Kings	C. P. R.
Oromocto lake	415	York	Geol. Surv.
Ortonville	352	Victoria	C. P. R.
Otter lake	345	St. John	Geol. Surv.
Oulton		Westmorland	
Station	72		N. B. & P. E. I. R.
Summit, 1·7 mile W	129		N. B. & P. E. I. R.
Painsec		Westmorland	
Station	149		I. C. R.
Summit, 2·7 miles S	180		I. C. R.
Palfrey lake	378	York	Coast Surv.
Palmerston	67	Kent	St. L. & R. R.
Paquetville	210	Gloucester	Geol. Surv.
Passekeag	33	Kings	I. C. R.
Peel	163	Carleton	C. P. R.
Pelerin settlement	266	Kent	Geol. Surv.
Peltoma settlement	550	York	Geol. Surv.
Pemberton ridge	750	York	Geol. Surv.
Pennfield	228	Charlotte	S. L. R.
Penniac	61	York	C. E. R.
Pennlyn	47	Queens	C. R. of N. B.
Penobsquis		Kings	
Station	93		I. C. R.
Salmon river, bed 70; rail	85		I. C. R.

NEW BRUNSWICK

Locality	Elev.	County	Authority
Perry settlement.	440	Kings	Geol. Surv.
Perth		Victoria	
Station	243		C. P. R.
St. John river, high water, 237 ; rail	246		C. P. R.
Petitcodiac		Westmorland	
Station	100		I. C. R.
Summit, 1·5 mile E	119		I. C. R.
Petite Roche station	83	Gloucester	I. C. R.
Piccadilly mountain	1,000	Kings	Geol. Surv.
Plaster Rock station	379	Victoria	C. P. R.
Pleasant lake	1,200	Kings	Geol. Surv.
Pleasant, Mount	1,175	Charlotte	Ganong
Pleasant ridge	590	Northumberland	Geol. Surv.
Plumweseep		Kings	
Station	67		I. C. R.
Penobsquis river, 1·0 mile W., bed 47 ; rail 65.	65		I. C. R.
Pointe du Chêne	9	Westmorland	I. C. R.
Pokemouche	11	Gloucester	Car. R.
Pokeshaw	62	Gloucester	Car. R.
Pole hill	935	Carleton	Geol. Surv.
Pollet lake	968	Kings	Geol. Surv.
Pollet River station	80	Westmorland	I. C. R.
Portage lake	1,298	Victoria	Ganong
Port Elgin	16	Westmorland	N. B. & P. E. I. R.
Prince William	465	York	C. P. R.
Quaker Brook station	275	Victoria	C. P. R.
Quispamsis		Kings	
Station	159		I. C. R.
Summit, 0·2 mile E	179		I. C. R.
Red Pine	338	Gloucester	I. C. R.
Red Rapids station	338	Victoria	C. P. R.
Reed Island station	370	Victoria	C. P. R.
Richibucto	20	Kent	St. L. & R. R.
Richmond corner	620	Carleton	Geol. Surv.
Riley mountain or Bald head	1,866	Victoria	Geol. Surv.
Riverbank (now Ononette)	50	Kings	C. P. R.
River de Chute station	279	Carleton	C. P. R.
Riverside	20	Kings	I. C. R.
Roach settlement	465	York	Geol. Surv.
Robertville	400	Gloucester	Geol. Surv.
Rockland	49	York	C. P. R.
Rogers head	400	St. John	Admir. chart
Rogersville	305	Northumberland	I. C. R.
Roix Road station	203	Charlotte	C. P. R.
Rolling Dam	233	Charlotte	C. P. R.
Rothesay	23	Kings	I. C. R.
Rowena	291	Victoria	C. P. R.
Rusagonis	38	Sunbury	C. P. R.
Sackville	26	Westmorland	I. C. R.
Saddleback settlement	1,000	Kings	Geol. Surv.
Sagamook or Bald mountain	2,604	Restigouche	Bdy. Com.
St. Andrews		Charlotte	
Station	23		C. P. R.
Wharf	12		C. P. R.
St. Anne	450	Madawaska	C. P. R.
St. Anthony (summit)¹	187	Kent	B. & M. R.
St. Basil	474	Madawaska	C. P. R.
St. Croix		York	
Station	387		C. P. R.
St. Croix river, high water 309 ; rail.	387		C. P. R.
St. George	87	Charlotte	S. L. R.
St. Hilaire	508	Madawaska	Tem. R.
St. Jacques	478	Madawaska	Tem. R.
St. John		St. John	
Station	17·5		I. C. R.
Wharf	15·5		I. C. R.
St. John river bridge, rail	85		C. P. R.
B. M. in foundation, S.E. cor. of custom house	41·63		Tidal Surv.
Zero of night gauge	21·59		Tidal Surv.

NEW BRUNSWICK

Locality	Elev.	County	Authority
St. John......		St. John.	
Extreme high tide (Oct. and Nov., 1896).........	+14·75		Tidal Surv.
Low water, spring tides (probably low water datum of chart, survey of 1887)...	−12·01		Tidal Surv.
Public Works datum	−12·69		Tidal Surv.
Tidal Survey datum	−13·95		Tidal Surv.
Harmonic tide plane	−12·53		Tidal Surv.
St. Joseph (summit)	102	Gloucester	Car. R.
St. Leonards	504	Madawaska	C. P. R.
St. Louis (formerly Palmerston)............	67	Kent	St. L. & R. R.
St. Louise	250	Gloucester	Geol. Surv.
St. Marys......	30	York	C. P. R.
St. Simon...........	30	Gloucester	Car. R.
St. Stephen.............	12	Charlotte	C. P. R.
Salamanca........	45	York	C. P. R.
Salisbury	102	Westmorland	I. C. R.
Salmon Beach.............	45	Gloucester	Car. R.
Sandy bay	378	York	Coast Surv.
Scotch settlement	250	Kent	Geol. Surv.
Scotch Settlement station............	120	Westmorland	B. & M. R.
Scotch settlement	600	Kings	Geol. Surv.
Scott	464	York	C. P. R.
Scott settlement	470	Carleton	Geol. Surv.
Serpentine lake	1,450	Northumberland	Geol. Surv.
Shediac	47	Westmorland	I. C. R.
Shepody mountain.........	1,050	Albert	Admir. chart
Shepody Road.............	850	Kings	Geol. Surv.
Shewen............		Carleton	
Station...	631		C. P. R.
Summit, 0·2 mile E........	641		C. P. R.
Shippegan	14	Gloucester	Car. R.
Shogomoc.................	226	York	C. P. R.
Siegas	442	Victoria	C. P. R.
Skedaddle ridge.......	1,425	Carleton	Geol. Surv.
Skiff lake	650	York	Geol. Surv.
Slate mountain...	2,000	Restigouche	Admir. chart
Smith corner........	70	York	C. P. R.
Snider mountain.....	720	Kings	Geol. Surv.
South Bay	37	St. John	C. P. R.
South Branch		Sunbury	
Station..	40		C. P. R.
Oromocto river, 0·9 mile W., high water, 31 ; low water, 14 ; rail.....	35		C. P. R.
South Knowlesville....	770	Carleton	Geol. Surv.
Spring Brook settlement	158	Kent	Geol. Surv.
Springfield hill....	700	York	Geol. Surv.
Springhill............	450	Kings	Geol. Surv.
Springhill station.............	68	York	C. P. R.
Springhill village...........	37	York	St. J. V. & R. L. R.
Spruce lake	207	St. John	S. L. R.
Squaw mountain	2,000	Restigouche	Admir. chart
Stanley.................	380	York	Geol. Surv.
Steeves mountain..............	373	Westmorland	Geol. Surv.
Steeves station	384	Westmorland	E. & H. R.
Stevens (now Acamac)..........	45	St. John	C. P. R.
Stonehaven.................		Gloucester	
Station	103		Car. R.
Summit, 1·0 mile E......	146		Car. R.
Stoneridge........	146	York	C. P. R.
Sugar Brook station............	466	York	C. P. R.
Sugarloaf mountain........	950	Restigouche	Admir. chart
Sussex...............	69	Kings	I. C. R.
Sutton (now Ketepec)	75	St. John	C. P. R.
Teed Mill..............	376	Carleton	C. P. R.
Teneriffe, Mount..........	2,108	Northumberland	Ganong
Third lake (Don branch of Tobique river)...........	1,248	Victoria	Geol. Surv.
Three Brooks station...	380	Victoria	C. P. R.
Three Tree Creek station.....	43	Sunbury	C. P. R.
Tiarks, Lake.....	1,517	Restigouche	I. C. R.

NEW BRUNSWICK

Locality	Elev.	County	Authority
Tilley settlement	650	Victoria	Geol. Surv.
Tobique Narrows station	301	Victoria	C. P. R.
Torryburn	75	St. John	I. C. R.
Tracadie	17	Gloucester	G. S. R.
Tracey	95	Sunbury	C. P. R.
Trousers or Tobique lake	1,243	Victoria	Geol. Surv.
Trout lake	1,445	Victoria	Ganong
Union corner	640	Carleton	Geol. Surv.
Upper Blackville	211	Northumberland	C. E. R.
Upper Caraquet	50	Gloucester	Car. R.
Upper Cross Creek station	324	York	C. E. R.
Upper Dorchester	27	Westmorland	I. C. R.
Upper Keswick	273	York	C. P. R.
Upper Knoxford	635	Carleton	Geol. Surv.
Upper Sackville	36	Westmorland	N. B. & P. E. I. R.
Upper Woodstock		Carleton	
Station	143		C. P. R.
St. John river, low water 99 ; rail	137		C. P. R.
Upsalquitch lake	862	Northumberland	Bdy. Com.
Victoria lake	495	Charlotte	Geol. Surv.
Victoria settlement	325	Sunbury	Geol. Surv.
Waasis	59	Sunbury	C. P. R.
Wadawamketch or Bald mountain	900	York	Geol. Surv.
Wanamake	650	Kings	Geol. Surv.
Washademoak	70	Queens	C. R. of N. B.
Waterford	640	Kings	Geol. Surv.
Waterville	315	Carleton	Geol. Surv.
Watson settlement	570	Carleton	Geol. Surv.
Watt	305	Charlotte	C. P. R.
Waweig	139	Charlotte	C. P. R.
Welsford	85	Queens	C. P. R.
Westfield Beach	25	Kings	C. P. R.
West St. John	15	St. John	C. P. R.
West Waweig	196	Charlotte	S. L. R.
Wheaton Mill	222	Westmorland	R. & H. R.
Whites mountain	790	Kings	Geol. Surv.
Wicklow	260	Carleton	Geol. Surv.
Williamsburg settlement	1,050	York	Geol. Surv.
Willow Grove	307	St. John	Geol. Surv.
Wood lake	522	St. John	Geol. Surv.
Woodstock		Carleton	
Station	134		C. P. R.
Queen street	136		C. P. R.
Meduxnakeag river, bed 99 ; rail	136		C. P. R.
Woodstock Road		York	
Station	546		C. P. R.
Summit near station	558		C. P. R.
Woolastook (formerly Ballentine)	33	Kings	C. P. R.
Young Cove station	216	Queens	C. R. of N. B.
Zealand	105	York	C. P. R.
Zionville	132	York	C. E. R.
Zionville settlement	620	York	Geol. Surv.

NOVA SCOTIA

Locality	Elev.	County	Authority
Afton.............		Antigonish	
Station......	96		I.C.R.
Summit, 0·4 mile E...	150		I.C.R.
Ainslie, Lake.........	188	Inverness	I. & R. R.
Alba.........	15	Inverness	I.C.R.
Albany...	573	Annapolis	C. R. of N. S.
Alexander...	216	Inverness	I. & R. R.
Alpena...........	534	Annapolis	C. R. of N. S.
Alton....		Colchester	
Station..........	86		I.C.R.
Summit, 1·4 mile N......	131		I.C.R.
Amherst...		Cumberland	
Station.....	63		I.C.R.
Summit, 1·0 mile S....	105		I.C.R.
Annapolis....	18	Annapolis	D.A.R.
Antigonish.........	20	Antigonish	I.C.R.
Arcadia...	23	Yarmouth	H. & Y. R.
Argyle...	10	Yarmouth	H. & Y. R.
Athol....	134	Cumberland	I.C.R.
Auburn..........	94	Kings	D.A.R.
Avondale...	151	Pictou	I.C.R.
Avonport........	57	Kings	D.A.R.
Aylesford....	100	Kings	D.A.R.
Barney River station...........	183	Pictou	I.C.R.
Barrachois...	21	Cape Breton	I.C.R.
Barrington Passage.......	10	Shelburne	H. & Y. R.
Bayfield Road......		Antigonish	
Station.......	135		I.C.R.
Summit, 0·8 mile W......	188		I.C.R.
Bear hill....	725	Guysborough	Admir. chart
Bear hill....	750	Inverness	Admir. chart
Bear River station...	30	Digby	D.A.R.
Beaverbank....	229	Halifax	D.A.R.
Bedford.....	44	Halifax	I.C.R.
Belleville.......	53	Yarmouth	H. & Y. R.
Belliveau....	151	Digby	D.A.R.
Belmont....	84	Colchester	I.C.R.
Berwick....	140	Kings	D.A.R.
Black River station....	204	Inverness	I. & R. R.
Black Rock siding...	257	Pictou	N. S. S. C. R.
Blockhouse......	180	Lunenburg	C. R. of N. S.
Blomidon, Cape......	570	Kings	Admir. chart
Bloomfield....		Digby	
Station......	309		D.A.R.
Summit, 0·5 mile W...	332		D.A.R.
Boisdale....	12	Cape Breton	I.C.R.
Boisdale hills....	700	Cape Breton	Admir. chart
Boss point...	210	Cumberland	Admir. chart
Brazil Lake station......	211	Yarmouth	D.A.R.
Bridgeport...	72	Cape Breton	S. & L. R.
Bridgetown....	29	Annapolis	D.A.R.
Bridgeville....	177	Pictou	N. S. S. C. R.
Bridgewater....	13	Lunenburg	C. R. of N. S.
Brierly Brook station....	148	Antigonish	I.C.R.
Broad Cove station....	52	Inverness	I. & R. R.
Brookfield....	102	Colchester	I.C.R.
Brooklyn....	33	Hants	M. R.
Browns Point....	11	Pictou	I.C.R.
Bucklaw hill...	952	Victoria	Admir. chart
Burton....		Hants	
Station......	141		M. R.
Summit, 1·4 mile W., rail..........	171		M. R.
Calder hill....	750	Inverness	Admir. chart
Caledonia mine....	83	Cape Breton	S. & L. R.
Cambridge....	71	Kings	D.A.R.
Cameron Mine siding...........	184	Pictou	N. S. S. C. R.

NOVA SCOTIA

Locality	Elev.	County	Authority
Campbell		Colchester	
Station	420		I. C. R.
Black river, bed 300 ; rail	406		I. C. R.
Canning	72	Kings	D. A. R.
Cape Dauphin hills	1,045	Victoria	Admir. chart
Cape Porcupine station	398	Antigonish	I. C. R.
Catalone	37	Cape Breton	S. & L. R.
Central Argyle	40	Yarmouth	H. & Y. R.
Centreville	90	Kings	D. A. R.
Chapman settlement	110	Cumberland	Geol. Surv.
Charcoal junction	191	Pictou	N. S. S. C. R.
Cherryfield	327	Lunenburg	C. R. of N. S.
Cheticamp hills		Inverness	
North summit	1,100		Admir. chart
South summit	1,130		Admir. chart
Chignecto isthmus, thalweg	66	Cumberland	C. M. T. R.
Church Point	147	Digby	D. A. R.
Claremont hill	565	Cumberland	Geol. Surv.
Clarksville		Hants	
Station	70		M. R.
Kennetcook river, 1·7 mile E., bed, 162 ; rail	177		M. R.
Clementsport	68	Annapolis	D. A. R.
Clifton	31	Colchester	M. R.
Cobequid mountains	990	Cumberland	Geol. Surv.
Coldbrook	71	Kings	D. A. R.
Conns Mills	25	Cumberland	I. C. R.
Cox hill	550	Cape Breton	Admir. chart
Coxheath hills	550	Cape Breton	Admir. chart
Craignish	82	Inverness	I. & R. R.
Craignish hills		Inverness	
East summit	980		Admir. chart
West summit	850		Admir. chart
Dalhousie		Annapolis	
Station	643		C. R. of N. S.
Summit, 0·3 mile N.	644		C. R. of N. S.
Nictaux river, 4·4 miles N., water 567 ; rail	577		C. R. of N. S.
Dalhousie or Fitzpatrick mountain	950	Pictou	Admir. chart
Dallas Brook hill	768	Inverness	Admir. chart
Dartmouth	14	Halifax	I. C. R.
Dartmouth lake	57	Halifax	Bouchette
Dawson	312	Hants	D. A. R.
Debert	151	Colchester	I. C. R.
Denmark	141	Colchester	I. C. R.
Digby	33	Digby	D. A. R.
Digby neck	425	Digby	Admir. chart
Dimock	30	Hants	M. R.
Doddridge	140	Hants	M. R.
Dominion	74	Cape Breton	S. & L. R.
East Mines		Colchester	
Station	193		I. C. R.
Folly river, 0·8 mile N., bed 109 ; rail	191		I. C. R.
East Pubnico	14	Yarmouth	H. & Y. R.
East Southampton	133	Cumberland	C. R. & C. Co.
Eastville	138	Colchester	M. R.
Eigg mountain	1,000	Antigonish	Admir. chart
Ellershouse	258	Hants	D. A. R.
Elmsdale	56	Hants	I. C. R.
Enfield		Hants	
Station	63		I. C. R.
Shubenacadie river, water 31 ; rail	56		I. C. R.
Eureka	145	Pictou	I. C. R.
Fairview	11	Halifax	I. C. R.
Falmouth	27	Hants	D. A. R.
Fenerty	266	Halifax	D. A. R.
Fenerty Lake siding	254	Halifax	D. A. R.
Ferrona junction	129	Pictou	I. C. R.
Ferrona siding	113	Pictou	N. S. S. C. R.
Ferry Road	54	Yarmouth	H. & Y. R.
Fletcher lake	51	Halifax	Bouchette

NOVA SCOTIA

Locality	Elev.	County	Authority
Folleigh (summit)	616	Colchester	I. C. R.
Folly lake	605	Colchester	I. C. R.
Fort Lawrence		Cumberland	
Siding	33		I. C. R.
Wharf	21·5		C. M. T. R.
Chignecto Marine Transport Ry (abandoned) crossing	33		C. M. T. R.
B. M. on masonry box culvert, N. side of I. C. R. track, at C. M. T. R. crossing	29·8		C. M. T. R.
B. M. at W. end of masonry box culvert, 2,120 feet S. of I. C. R. crossing	26·4		C. M. T. R.
Saxby tide (Oct. 5, 1869), highest known	29·2		C. M. T. R.
Gardiner	77	Cape Breton	S. & L. R.
Gasperau lake	46	Halifax	I. C. R.
George, Cape	310	Antigonish	Admir. chart
George mountain	890	Cape Breton	Admir. chart
George River station	37	Cape Breton	I. C. R.
Glace Bay station	72	Cape Breton	S. & L. R.
Glen Dhu or Alexander		Inverness	
Station	216		I. & R. R.
Summit, 1·0 mile S	244		I. & R. R.
Glengarry		Inverness	
Station	65		I. & R. R.
Summit, 1·6 mile S	110		I. & R. R.
Glengarry	395	Pictou	I. C. R.
Gordon Summit siding	559	Pictou	I. C. R.
Grand lake	46	Halifax	I. C. R.
Grand Lake station	58	Halifax	I. C. R.
Grand Lake station	104	Cape Breton	S. & L. R.
Grand Narrows	12	Cape Breton	I. C. R.
Grand Pré	28	Kings	D. A. R.
Green Oaks	68	Colchester	M. R.
Greenville	200	Cumberland	I. C. R.
Grove	349	Hants	D. A. R.
Halifax		Halifax	
Station	57		I. C. R.
I. C. wharf	14		I. C. R.
High service reservoir, waste weir	360		City Eng.
Low service reservoir, waste weir	204·5		City Eng.
Highest point within city limits	249		City Eng.
Citadel	227		Admir. chart
B. M., near S. E. angle of sail loft at dockyard	12·71		Tidal Surv.
Coping of dry dock	+ 7·60		Tidal Surv.
Harmonic tide plane	− 2·96		Tidal Surv.
Admiralty datum (low water, ordinary spring tides)	− 3·37		Tidal Surv.
Tidal Survey datum	− 2·36		Tidal Surv.
Sill of dry dock	−28·86		Tidal Surv.
Hansford hill	295	Cumberland	Geol. Surv.
Hantsport	48	Hants	D. A. R.
Harbour au Bouche	271	Antigonish	I. C. R.
Hastings	8	Inverness	I. & R. R.
Haute, Isle	320	Cumberland	Admir. chart
Hawkesbury	28	Richmond	I. C. R.
Hay and Hibbert	466	Hants	D. A. R.
Heatherton	20	Antigonish	I. C. R.
Hebron	82	Yarmouth	D. A. R.
Hectanooga	163	Digby	D. A. R.
Hemford	330	Queens	N. S. S. C. R.
High capes	1,000	Inverness	Admir. chart
Hilden		Colchester	
Station	141		I. C. R.
Summit, 1·4 mile S., ground 187; rail	181		I. C. R.
Hillaton	30	Kings	D. A. R.
Hingley	335	Colchester	I. C. R.
Homeville	94	Cape Breton	S. & L. R.
Hopewell	206	Pictou	I. C. R.
Horton Landing	44	Kings	D. A. R.
Hub mine	120	Cape Breton	S. & L. R.
Hunter mountain	680	Victoria	Admir. chart
Ingonish mountain	1,392	Victoria	Admir. chart

NOVA SCOTIA

Locality	Elev.	County	Authority
Iona	12	Victoria	I. C. R.
James River		Antigonish	
Station	255		I. C. R.
James river, water 188; rail	203		I. C. R.
Jeddore head	200	Halifax	Admir. chart
Jeffers	82	Cumberland	C. R. & C. Co.
Joggins	56	Cumberland	C. C. & R. Co.
Jordantown	165	Digby	D. A. R.
Judique	8	Inverness	I. & R. R.
Kennetcook		Hants	
Station	97		M. R.
Kennetcook river, 1·7 mile E., bed 89; rail	104		M. R.
Kentville	38	Kings	D. A. R.
Kingsport	47	Kings	D. A. R.
Kingston	85	Kings	D. A. R.
Lake Annis station	160	Yarmouth	D. A. R.
Lakeland	67	Cumberland	C. R. & C. Co.
Lakeview	137	Halifax	I. C. R.
Lansburg	505	Pictou	I. C. R.
Lansdowne	463	Pictou	I. C. R.
Lawrence	115	Cumberland	C. R. & C. Co.
Lawrencetown	75	Annapolis	D. A. R.
Leitche Creek station	10	Cape Breton	I. C. R.
Lily lake	134	Halifax	I. C. R.
Lily Lake station	141	Halifax	I. C. R.
Lime Quarry siding	164	Pictou	N. S. S. C. R.
Linwood	127	Antigonish	I. C. R.
Little Brook station	140	Digby	D. A. R.
Little Judique	6	Inverness	I. & R. R.
Little River village	350	Cumberland	Geol. Surv.
Lochbroom	10	Pictou	I. C. R.
Londonderry	334	Colchester	I. C. R.
Long island	210	Digby	Admir. chart
Long lake	86	Halifax	I. C. R.
Long Point hills	700	Inverness	Admir. chart
Long Point station	39	Inverness	I. & R. R.
Lorne		Pictou	
Siding	365		I. C. R.
Middle river of Pictou, 1·1 mile W., bed 309; rail	375		I. C. R.
Louisburg		Cape Breton	
Station	54		S. & L. R.
Shipping pier	42		S. & L. R.
Lourdes	18	Pictou	I. C. R.
Lower Argyle	66	Yarmouth	H. & Y. R.
Lower East Pubnico	11	Yarmouth	H. & Y. R.
Lower Wentworth village	160	Cumberland	Geol. Surv.
Lunenburg		Lunenburg	
Station	30		C. R. of N. S.
Wharf	9		C. R. of N. S.
Mabou	46	Inverness	I. & R. R.
Mabou highlands	1,000	Inverness	Admir. chart
Maccan	31	Cumberland	I. C. R.
Mackay point	610	Victoria	Admir. chart
Mahone	84	Lunenburg	C. R. of N. S.
Maitland		Lunenburg	
Station	204		C. R. of N. S.
Summit, 0·3 mile S	230		C. R. of N. S.
Malagash (Ross Road)	48	Cumberland	I. C. R.
Mapleton	133	Cumberland	C. R. & C. Co.
Marsh Point hills	560	Inverness	Admir. chart
Marshy Hope		Pictou	
Station	373		I. C. R.
Summit, 1·0 mile E	451		I. C. R.
McDonald Mine siding	180	Pictou	N. S. S. C. R.
McIntyre Lake station	133	Richmond	I. C. R.
McKinnon Harbour station	9	Victoria	I. C. R.
McMullin	42	Hants	M. R.
Meadowdale	31	Hants	M. R.
Meadowville	141	Pictou	I. C. R.

NOVA SCOTIA

Locality	Elev.	County	Authority
Merigomish......	18	Pictou	I. C. R.
Meteghan....	125	Digby	D. A. R.
Middle Bridge........		Inverness	
Station..	213		I. & R. R.
Summit, 0·9 mile W., ground 234 ; rail............	221		I. & R. R.
Middle Stewiacke	80	Colchester	M. R. of N. S.
Middleton.........	73	Annapolis	D. A. R.
Milford.....		Hants	
Station	52		I. C. R.
Summit, 1·7 mile S	101		I. C. R.
Millbrook	37	Hants	M. R.
Mill Village............................	45	Kings	D. A. R.
Mines Road	142	Richmond	I. C. R.
Mira...........................	12	Cape Breton	S. & L. R.
Monastery	13	Antigonish	I. C. R.
Morien junction	124	Cape Breton	S. & L. R.
Morien station..................	164	Cape Breton	S. & L. R.
Mount Uniacke		Hants	
Station	509		D. A. R.
Summit, 0·1 mile S	514		D. A. R.
Mulgrave............	9	Guysborough	I. C. R.
Nappan......	20	Cumberland	I. C. R.
New Germany......	260	Lunenburg	C. R. of N. S.
New Glasgow......	29	Pictou	I. C. R.
Newport	119	Hants	D. A. R.
Newville	66	Cumberland	C. R. & C. Co.
Nictaux......		Annapolis	
Station	150		C. R. of N. S.
Annapolis river, 2·6 miles N., bed 17 ; rail........	59		C. R. of N. S.
North cape..................	1,000	Victoria	Admir. chart
Northfield	133	Lunenburg	C. R. of N. S.
North mountain	700	Annapolis	Admir. chart
North Range	302	Digby	D. A. R.
North Sydney junction................	159	Cape Breton	I. C. R.
North Sydney station	41	Cape Breton	I. C. R.
Oakfield	62	Halifax	I. C. R.
Ohio	157	Yarmouth	D. A. R.
Old Tank......	77	Cape Breton	S. & L. R.
Orangedale........	12	Inverness	I. C. R.
Oxford junction	94	Cumberland	I. C. R.
Oxford station	38	Cumberland	I. C. R.
Paradise......................	50	Annapolis	D. A. R.
Parrsboro	21	Cumberland	C. R. & C. Co.
Pictou....	9	Pictou	I. C. R.
Pictou Landing....	10	Pictou	I. C. R.
Piedmont........		Pictou	
Station	241		I. C. R.
Summit, 1·2 mile E.....	288		I. C. R.
Pine Tree.... ...	16	Pictou	I. C. R.
Pitman Road....	231	Yarmouth	D. A. R.
Pleasant Lake station	16	Yarmouth	H. & Y. R.
Plympton........	292	Digby	D. A. R.
Point Tupper	9	Richmond	I. C. R.
Pomquet	43	Antigonish	I. C. R.
Port Clyde	26	Shelburne	H. & Y. R.
Port Hood	65	Inverness	I. & R. R·
Port Williams.................	31	Kings	D. A. R.
Prim point....	400	Digby	Admir. chart
Princeport Road.		Colchester	
Station.....................	212		M. R.
Summit, 0·2 mile W........	233		M. R.
Pubnico	10	Yarmouth	H. & Y. R.
Pugwash junction.....		Cumberland	
Station	37		I. C. R.
Summit, 1·1 mile W	73		I. C. R.
Pugwash station....	14	Cumberland	I. C. R.
Red cape	900	Inverness	Admir. chart
Red head................	272	Victoria	Admir. chart
Red head.................	500	Colchester	Admir. chart

NOVA SCOTIA

Locality	Elev.	County	Authority
Reserve mine	172	Cape Breton	S. & L. R.
Richmond	15	Halifax	I. C. R.
Ridge Road station	576	Annapolis	C. R. of N. S.
River Denys station	72	Inverness	I. C. R.
River Hebert station	29	Cumberland	C. C. & R. Co.
River John station	67	Pictou	I. C. R.
River Philip		Cumberland	
Station (summit)	172		I. C. R.
River Philip, 2·1 miles E., bed 41; rail	95		I. C. R.
Riverdale	125	Lunenburg	C. R. of N. S.
Riverdale	314	Colchester	I. C. R.
Riverside	20	Kings	I. C. R.
Riverton	120	Pictou	I. C. R.
Rockingham	18	Halifax	I. C. R.
Rocky Lake station	134	Halifax	I. C. R.
Rodney corner	625	Cumberland	Geol. Surv.
Roundhill	32	Annapolis	D. A. R.
St. Anne mountain	1,070	Victoria	Admir. chart
St. Croix	35	Hants	M. R.
St. Paul	206	Pictou	N. S. S. C. R.
St. Paul island		Victoria	
North summit	400		Admir. chart
South summit	500		Admir. chart
Salem village	450	Cumberland	Geol. Surv.
Salt hill	717	Inverness	Admir. chart
Saltsprings	153	Cumberland	I. C. R.
Sand Point station	16	Shelburne	N. S. S. R.
Sandy Cove station	62	Halifax	I. C. R.
Saulnierville	117	Digby	D. A. R.
Scotch Hill station	61	Pictou	I. C. R.
Scotch Village	133	Hants	M. R.
Scotsburn	251	Pictou	I. C. R.
Shag Harbour station	59	Shelburne	H. & Y. R.
Sheffield Mills	92	Kings	D. A. R.
Shelburne	69	Shelburne	N. S. S. R.
Shenacadie	14	Cape Breton	I. C. R.
Shenecadie hill	670	Cape Breton	Admir. chart
Shubenacadie		Hants	
Station	66		I. C. R.
Shubenacadie river, water 22; rail	55		I. C. R.
Shubenacadie lake, First	93	Halifax	Bouchette
Shubenacadie lake, Great	38	Halifax	Bouchette
Shubenacadie lake, Second	63	Halifax	Bouchette
Sissiboo Falls	288	Digby	D. A. R.
Smith Cove station	44	Digby	D. A. R.
Smoke cape	950	Victoria	Admir. chart
Southampton	112	Cumberland	C. R. & C. R.
South Brookfield	263	Queens	N. S. S. R.
South Maitland		Hants	
Station	32		M. R.
Shubenacadie river, extreme high water, 29; rail	44		M. R.
South mountain	530	Annapolis	Admir. chart
South River station	20	Antigonish	I. C. R.
South Uniacke		Hants	
Siding	449		D. A. R.
Summit, 1·0 mile S.	456		D. A. R.
Southwest Mabou	83	Inverness	I. & R. R.
Split cape	400	Kings	Admir. chart
Springfield	551	Annapolis	C. R. of N. S.
Springhill junction		Cumberland	
Station	199		I. C. R.
Summit, 1·8 mile S.	274		I. C. R.
Springhill Mines		Cumberland	
Station	448		C. R. & C. Co.
Summit	610		Geol. Surv.
Springhill Jct. switch	491		C. R. & C. Co.
Springside	135	Colchester	M. R.
Springville siding	182	Pictou	N. S. S. C. R.
Squirrel mountain	1,220	Inverness	Admir. chart

NOVA SCOTIA

Locality	Elev.	County	Authority
Stanley....	38	Hants	M. R.
Stellarton.	58	Pictou	I. C. R.
Stewiacke.		Colchester	
Station	86		I. C. R.
Stewiacke river, 1·8 mile N., water 24 ; rail..	44		I. C. R.
Stewiacke Cross Roads P. O..	134	Colchester	M. R.
Stillwater.	412	Hants	D. A. R.
Strath Lorne.		Inverness	
Station	115		I. & R. R.
Summit, 2·1 miles W., ground 204 ; rail.	199		I. & R. R.
Sugarloaf hill.	760	Antigonish	Admir. chart
Sugarloaf hill (Malignant brook).	690	Antigonish	Admir. chart
Sunnybrae.	207	Pictou	N. S. S. C. R.
Sutherland lake.	730	Cumberland	Geol. Surv.
Sydney		Cape Breton	
Station	8		I. C. R.
Shipping pier	42		S. & L. R.
Sylvester.	20	Pictou	I. C. R.
Tatamagouche.	15	Colchester	I. C. R.
Taylor Road..	113	Antigonish	I. C. R.
Thomson.	107	Cumberland	I. C. R.
Three-mile Plains.	79	Hants	D. A. R.
Tracadie.	48	Antigonish	I. C. R.
Trenton.	41	Pictou	I. C. R.
Truro .		Colchester	
Station .	60		I. C. R.
Salmon river, bed 35 ; rail..	51		I. C. R.
Tusket.	54	Yarmouth	H. & Y. R.
Union	218	Colchester	I. C. R.
Upper Stewiacke	90	Colchester	M. R. of N. S.
Upper Woods Harbour.	10	Shelburne	H. & Y. R.
Valley.	102	Colchester	I. C. R.
Wallace.	132	Cumberland	I. C. R.
Wallace Bridge	48	Cumberland	I. C. R.
Warren	225	Cumberland	Geol. Surv.
Waterville.	93	Kings	D. A. R.
Waverley siding (main line).	138	Halifax	I. C. R.
Waverley station (Dartmouth branch).	70	Halifax	I. C. R.
Wellington.	76	Halifax	I. C. R.
Wentworth station	472	Cumberland	I. C. R.
Wentworth village	210	Cumberland	Geol. Surv.
West Bay Road		Inverness	
Station	214		I. C. R.
Summit, 2·2 miles N., ground 298 ; rail.	286		I. C. R.
Westbrook	108	Cumberland	C. R. & C. Co.
Westchester station	299	Cumberland	I. C. R.
Westchester village	745	Cumberland	Geol. Surv.
West Merigomish	65	Pictou	I. C. R.
West River		Pictou	
Station	441		I. C. R.
West river of Pictou, bed 419 ; rail.	427		I. C. R.
Westville.	216	Pictou	I. C. R.
Weymouth	53	Digby	D. A. R.
White caps.	850	Inverness	Admir. chart
Wilkie Sugarloaf mountain	1,200	Victoria	Admir. chart
Wilkins..	26	Hants	D. A. R.
Wilmot.	70	Annapolis	D. A. R.
Windham hill	625	Cumberland	Geol. Surv.
Windsor.	26	Hants	D. A. R.
Windsor junction	129	Halifax	I. C. R.
Wolfville.	28	Kings	D. A. R.
Woodburn		Pictou	
Station	136		I. C. R.
Summit, 2·7 miles S.	218		I. C. R.
Woods Harbour.	36	Shelburne	H. & Y. R.
Yarmouth			
Dom. Atl. station.	36		D. A. R.
Hal. & Yar. station.	12		H. & Y. R.

ONTARIO

Locality	Elev.	County	Authority
Aberarder	664	Lambton	G. T. R.
Abitibi lake	830	Nipissing	Geol. Surv.
Abiwin	1,437	Rainy River	C. N. R.
Acton West	1,198	Halton	G. T. R.
Addie lake	1,605	Thunder Bay	C. N. R.
Admaston		Renfrew	
Can. Pac. station	402		C. P. R.
Can. Atl. station	412		C. A. R.
King. & Pem. station	400		K. & P. R.
Agincourt		York	
Can. Pac. station	563		C. P. R.
Grand Trunk station	567		G. T. R.
Grand Trunk Ry. crossing; G. T. Ry. rail 544; C. P. Ry. rail	568		C. P. R.
Agitamo lake	1,042	Thunder Bay	C. P. R.
Ahmic lake	908	Parry Sound	Geol. Surv.
Ailsa Craig		Middlesex	
Station	753		G. T. R.
River Ausable, water 716; rail	749		G. T. R.
Aird island	773	Algoma	Admir. chart
Alexandria	256	Glengarry	C. A. R.
Alfred	178	Prescott	C. P. R.
Algoma	602	Algoma	C. P. R.
Allack lake	649	Renfrew	K. & P. R.
Allanburg station	591	Welland	G. T. R.
Allanburg junction	610	Welland	G. T. R.
Allandale	737	Simcoe	G. T. R.
Alliston	723	Simcoe	G. T. R.
Allumette or Pembroke lake, high water, 370; low water	363	Renfrew	Rys. & Canals
Alma	1,438	Wellington	G. T. R.
Almonte		Lanark	
Station	399		C. P. R.
Mississippi river, water 368; rail	402		C. P. R.
Alton	1,313	Peel	C. P. R.
Alvinston		Lambton	
Grand Trunk station	729		G. T. R.
Mich. Cent. station	713		M. C. R.
Amable du Fond lake	1,191	Nipissing	C. P. R.
Amaranth	1,544	Dufferin	C. P. R.
Amherstburg	593	Essex	M. C. R.
Amyot	1,386	Thunder Bay	C. P. R.
Ancaster	328	Wentworth	T. H. & B. R.
Angus	627	Simcoe	G. T. R.
Angus lake	1,051	Nipissing	Geol. Surv.
Anjigama lake	1,174	Algoma	C. P. R.
Annima-Nipissing lake	1,070	Nipissing	Geol. Surv.
Anson		Hastings	
Station	399		C. O. R.
Summit, 1·8 mile, N	455		C. O. R.
Antoine lake	900	Frontenac	K. & P. R.
Appin		Middlesex	
Grand Trunk station	740		G. T. R.
Can. Pac. station	747		C. P. R.
Mich. Cent. crossing	742		G. T. R.
Apple Hill	298	Glengarry	C. P. R.
Appleton (closed)	426	Lanark	C. P. R.
Ardendale		Frontenac	
Station	617		C. P. R.
Salmon river, bed, 609; rail	630		C. P. R.
Argyle	870	Victoria	G. T. R.
Arkell		Wellington	
Station	1,112		C. P. R.
Summit, 1·1 mile, S	1,167		C. P. R.

ONTARIO

Locality	Elev.	County	Authority
Arkwood	605	Kent	C. P. R.
Armstrong lake	580	Frontenac	K. & P. R.
Arner	602	Essex	L. E. & D. R. R.
Arnprior		Renfrew	
Can. Pac. station	301		C. P. R.
Can. Atl. station	302		C. A. R.
Madawaska river, high water, 259; low water, 242; rail	287		C. P. R.
Can. Pac. crossing	302		C. A. R.
Arrow lake	1,510	Thunder Bay	Geol. Surv.
Arthur	1,523	Wellington	C. P. R.
Ash (formerly Zimmerman)	573	Halton	G. T. R.
Ashdod	572	Renfrew	K. & P. R.
Ashton	446	Carleton	C. P. R.
Asylum		Middlesex	
Station	854		C. P.
Grand Trunk Ry. crossing	858		C. P.
Atbara	1,540	Thunder Bay	C. P. R.
Athens (formerly Farmersville)	416	Leeds	B. & W. R.
Atherly	727	Ontario	G. T. R.
Atikokan	1,290	Rainy River	C. N. R.
Atlantic and Northwest junction	404	Renfrew	C. P. R.
Attercliffe	588	Haldimand	M. C. R.
Atukamamoan lake		Rainy River	C. P. R.
Atwood	1,180	Perth	G. T. R.
Auburn Mills	1,098	Peterborough	G. T. R.
Aultsville		Stormont	
Grand Trunk station	248		G. T. R.
B. M., south coping of culvert ½ mile W. of station	239·5		D. W. Com.
B. M., west wall of small bridge ¾ mile W. of station	242·5		D. W. Com.
Aurora	886	York	G. T. R.
Avening	779	Simcoe	G. T. R.
Avonmore		Stormont	
Station	324		C. P. R.
Payne river, bed, 296; rail	306		C. P. R.
Aylen		Nipissing	
Station	725		C. P. R.
Summit, 1·4 mile S	733		C. P. R.
Aylen Lake		Nipissing	
Station	1,162		C. A. R.
Summit, 1·1 mile E., ground, 1,206; rail	1,184		C. A. R.
Aylmer		Elgin	
Grand Trunk station	761		G. T. R.
Mich. Cent. station	818		M. C. R.
Ayr		Waterloo	
Station	961		C. P. R.
River Nith, bed, 916; rail	950		C. P. R.
Baden	1,156	Waterloo	G. T. R.
Badgley island	950	Algoma	Admir. chart
Bainsville		Glengarry	
Station	174		G. T. R.
B. M., east abutment of beam culvert, near station	176		D. W. Com.
Baird	753	Elgin	G. T. R.
Baldwin	743	York	G. T. R.
Ballantrae		York	
Station	1,108		G. T. R.
Summit, 2·1 miles, S	1,121		G. T. R.
Ballantyne		Frontenac	
Station	360		G. T. R.
B. M., coping of box culvert, quarter mile, W	358		D. W. Com.
Ballsville	692	Haldimand	G. T. R.
Balsam lake	838	Victoria	Rys. & Canals
Baltimore		Northumberland	
Village	521		C. N. & P. R.
Mill-pond, water	523		C. N. & P. R.
Bancroft		Hastings	
Station	1,073		C. O. R.
York river, water above dam	1,067		C. O. R.
Bannerman lake	1,270	Algoma	Geol. Surv.

ONTARIO

Locality	Elev.	County	Authority
Banning	1,259	Rainy River	C. N. R.
Bannockburn		Hastings	
Station	828		C. O. R.
Moira river, bed, 836 ; rail	844		C. O. R.
Baptiste lake	1,205	Hastings	I. B. & O. R.
Baptiste Lake station	1,215	Hastings	I. B. & O. R.
Barclay	1,251	Rainy River	C. P. R.
Baril lake	1,496	Thunder Bay	Dawson
Bark lake	1,010	Renfrew	Geol. Surv.
Bark lake	1,522	Algoma	C. P. R.
Barlow lake	653	Nipissing	Geol. Surv.
Barrel lake	1,312	Rainy River	C. P. R.
Barrie	726	Simcoe	G. T. R.
Barrys Bay	988	Renfrew	C. A. R.
Barryvale	535	Renfrew	K. & P. R.
Bartonville	361	Wentworth	T. H. & B. R.
Barwick	1,090	Rainy River	C. N. R.
Base-line lake, high water	1,406	Rainy River	C. N. R.
Basket lake	1,337	Rainy River	C. P. R.
Bass lake	609	Renfrew	C. P. R.
Bass lake	724	Parry Sound	C. P. R.
Bass or Mountain lake	1,332	Renfrew	K. & P. R.
Basswood lake	1,290	Rainy River	Dawson
Bathurst	479	Lanark	C. P. R.
Batteaux	684	Simcoe	G. T. R.
Battery bluff	819	Algoma	Admir. chart
Battery lake	1,568	Thunder Bay	C. P. R.
Bay lake	800	Nipissing	Geol. Surv.
Bays, Lake of	1,038	Muskoka	Geol. Surv.
Beach Road	255	Wentworth	G. T. R.
Beachville		Oxford	
Can. Pac. station	918		C. P. R.
Grand Trunk station	907		G. T. R.
Beamsville	297	Lincoln	G. T. R.
Bearbrook		Russell	
Station	249		C. A. R.
Summit, 1·4 mile, W	254		C. A. R.
Bear lake	997	Nipissing	Geol. Surv.
Bear lake	1,183	Rainy River	C. P. R.
Bear lake	1,030	Parry Sound	C. A. R.
Bear Lake station	1,037	Parry Sound	C. A. R.
Bear Pass station	1,147	Rainy River	C. N. R.
Beatty	670	Halton	C. P. R.
Beatty siding (summit)	905	Parry Sound	C. A. R.
Beaucage	673	Nipissing	C. P. R.
Beaver lake	843	Frontenac	K. & P. R.
Beaver lake	881	Lanark	K. & P. R.
Beaver mountain	1,247	Nipissing	Geol. Surv.
Beaver P. O.	936	Thunder Bay	Geol. Surv.
Beaverton	761	Ontario	G. T. R.
Beckwith	461	Lanark	C. P. R.
Beckwith island	897	Simcoe	Admir. chart
Bedford		Frontenac	
Station	500		K. & P. R.
Glendower Mines branch, end of track	520		K. & P. R.
Summit, 1·0 mile, S	521		K. & P. R.
Beeton station	727	Simcoe	G. T. R.
Beeton junction		Simcoe	
Station	716		G. T. R.
Nottawasaga river 1·4 mile, N., bed, 680 ; rail	705		G. T. R.
Belgrave	1,058	Huron	G. T. R.
Bell	388	Leeds	C. P. R.
Bellamy	393	Leeds	C. P. R.
Bell Corners	299	Carleton	C. P. R.
Belle River		Essex	
Grand Trunk station	583		G. T. R.
Can. Pac. station	587		C. P. R.
Belleville, city station	251	Hastings	G. T. R.
Belleville junction	295	Hastings	G. T. R.

ONTARIO

Locality	Elev.	County	Authority
Belmont	844	Middlesex	C. P. R.
Belwood	1,415	Wellington	C. P. R.
Benallen	799	Grey	G. T. R.
Berkeley	1,330	Grey	C. P. R.
Berlin junction	1,004	Waterloo	G. T. R.
Berlin station	1,100	Waterloo	G. T. R.
Berwick	243	Stormont	N. Y. & O. R.,
Best		Victoria	
Station	886		G. T. R.
Summit, 1·6 mile, W	936		G. T. R.
Bethany	862	Durham	G. T. R.
Beverly lake, Lower	295	Leeds	Rys. & Canals
Bigby	1,488	Rainy River	C. P. R.
Big Whitefish lake	1,010	Nipissing	Geol. Surv.
Biota	1,553	Thunder Bay	C. P. R.
Birch lake	449	Frontenac	K. & P. R.
Birch lake	1,366	Thunder Bay	C. P. R.
Birch lake	1,442	Thunder Bay	Dawson
Bird Creek	1,108	Hastings	C. O. R.
Birdsall		Peterborough	
Station	617		G. T. R.
Ouse river, water, 613; rail	623		G. T. R.
Biskotasing		Algoma	
Station	1,388		C. P. R.
Summit, 4·3 miles, W., ground, 1414; rail	1,411		C. P. R.
Biskotasing lake	1,321	Algoma	C. P. R.
Bismarck	708	Elgin	M. C. R.
Bissett	550	Renfrew	C. P. R.
Blackburn	230	Carleton	C. P. R.
Black Creek station	574	Welland	M. C. R.
Blacks		Stormont	
Station	253		N. Y. & O. R.
Black river, bed, 243; rail	252		N. Y. & O. R.
Black River station	894	Thunder Bay	C. P. R.
Black Sturgeon lake	821	Thunder Bay	C. P. R.
Blackwater	859	Ontario	G. T. R.
Blackwell	601	Lambton	G. T. R.
Blair	908	Waterloo	G. T. R.
Blairton	643	Peterborough	C. P. R.
Blandford	971	Oxford	C. P. R.
Blenheim		Kent	
E. & H. station	681		L. E. & D. R. R.
L. E. & D. R. station	672		L. E. & D. R. R.
Blenheim junction		Kent	
Station	674		L. E. & D. R. R.
Summit, 9·1 mile, S., ground, 698; rail	683		L. E. & D. R. R.
Blezard	712	Peterborough	G. T. R.
Blind River station	603	Algoma	C. P. R.
Bline lake	910	Frontenac	K. & P. R.
Bloomfield	260	Prince Edward	C. O. R.
Blue mountains	1,635	Grey	Admir. chart
Bluevale	1,078	Huron	G. T. R.
Blyth		Huron	
Station	1,080		G. T. R.
Summit, 1·6 mile, N., ground, 1126; rail	1,121		G. T. R.
Blytheswood	629	Essex	M. C. R.
Boba lake	532	Frontenac	K. & P. R.
Bogie lake	1,007	Nipissing	Geol. Surv.
Boice lake	956	Nipissing	Geol. Surv.
Bolckow	1,381	Algoma	C. P. R.
Bolton	835	Peel	C. P. R.
Bolton Brook siding	686	Frontenac	K. & P. R.
Bonfield	782	Nipissing	C. P. R.
Bonheur		Rainy River	
Station	1,531		C. P. R.
Summit, 2·4 miles, W.	1,555		C. P. R.
Bothwell	688	Kent	G. T. R.
Bowmanville	261	Durham	G. T. R.
Bow Park	659	Brant	T. H. & B. R.

5½

ONTARIO

Locality	Elev.	County	Authority
Boyer lake........	1,260	Algoma	A. C. & H. B. R.
Bracebridge...	817	Muskoka	G. T. R.
Bradford..........	728	Simcoe	G. T. R.
Braeside..........	279	Renfrew	C. P. R.
Bramley...........	887	Simcoe	G. T. R.
Brampton........		Peel	
Grand Trunk station.............	712		G. T. R.
Can. Pac. station........	720		C. P. R.
Etobicoke river, water, 689 ; rail ..	712		G. T. R.
Grand Trunk Ry. crossing...	713		C. P. R.
Branchton................................	895	Waterloo	G. T. R.
Brandy Creek station.......	788	Norfolk	G. T. R.
Brant House	268	Halton	G. T. R.
Brantford......		Brant	
Grand Trunk (B. & G.) static	705		G. T. R.
Grand Trunk (B. & T.) station......... ..	666		G. T. R.
T. H. & B. station..........	657		T. H. & B. R.
Starch Works siding....	659		T. H. & B. R.
B. & G. and B. & T. crossing.......	687		G. T. R.
City datum—100 feet below top of foundation stone of building, S. E. corner Colborne and Market sts.	589		City Eng.
B. M., water table, Stratford's building S. W. corner, King and Dalhousie streets..	700		City Eng.
B. M., water table, engineer's house, pumping station.......................	668		City Eng.
City hall, ground floor......	691		City Eng.
Dundas st., near Sydenham st.......	786		City Eng.
Grand river—.			
Above Electric Co.'s dam, low water, 650 ; extreme high water (March, 1897)...	661		City Eng.
Below Electric Co.'s dam, low water, 641; extreme high water, (April 1900).............	657		City Eng.
Breadalbane lake...........	986	Nipissing	Geol. Surv.
Brechin..................................	745	Ontario	G. T. R.
Breeches lake	1,085	Nipissing	Geol. Surv.
Bremner		Thunder Bay	
Station..................	1,129		C. P. R.
White river, 2·9 miles, W., water, 1,107 ; rail.....	1,119		C. P. R.
Brentwood................................	646	Simcoe	G. T. R.
Breslau..................................		Waterloo	
Station....	1,023		G. T. R.
Grand river, water, 961 ; rail	1,010		G. T. R.
Bridgeburg (Victoria).	587	Welland	G. T. R.
Bridge River station (closed).	1,543	Thunder Bay	C. P. R.
Brigden.................................		Lambton	
Station..........................	634		M. C. R.
Bear creek, bed, 588 ; rail...........	626		M. C. R.
Bright..................................	1,040	Oxford	G. T. R.
Brighton...........	303	Northumberland	G. T. R.
Brinklow.............	1,152	Hastings	C. O. R.
Britannia	199	Carleton	C. P. R.
Britomart point............................	772	Algoma	Admir. chart
Broad hill...............................	1,233	Algoma	Admir. chart
Brockville		Leeds	
Grand Trunk station	280		G. T. R.
B. W. & S. Ste. M. station............	257		B. & W. R.
Brockville, Westport & S. Ste. M. Ry. crossing, B. W. & S. Ste. M. Ry., rail 306 ; G. T. Ry., rail.	282		G. T. R.
Wharf......	252		C. P. R.
B. M., coping of G. T. Ry. bridge over C. P. Ry., east of station.................	291		D. W. Com.
B. M., coping of G. T. Ry. bridge over C. P. Ry., west of station	277		D. W. Com.
Bronson Road.	1,077	Hastings	C. O. R.
Bronte..............	344	Halton	G. T. R.
Brookfield........		Welland	
Grand Trunk station....	593		G. T. R.
Mich. Cent. station	608		M. C. R.
Brookholm....	598	Grey	G. T. R.

ONTARIO

Locality	Elev.	County	Authority
Brooklin	538	Ontario	G. T. R.
Brophy lake	1,056	Nipissing	Geol. Surv.
Brown Hill station	754	York	G. T. R.
Brownsville	793	Oxford	M. C. R.
Brucefield	886	Huron	G. T. R.
Bruce Mines	683	Algoma	C. P. R.
Brulé	1,355	Rainy River	C. P. R.
Brulé Harbour hill	1,300	Algoma	Admir. chart
Brulé lake	872	Frontenac	C. P. R.
Brulé lake	1,389	Rainy River	C. P. R.
Brulé Lake		Nipissing	
Station	1,472		C. A. R.
Summit, 3·0 miles, W.	1,607		C. A. R.
Brunel	1,410	Algoma	C. P. R.
Brunswick	865	Durham	G. T. R.
Brussels	1,121	Huron	G. T. R.
Buckhorn lake	805	Peterborough	Rys. & Canals
Buckshot lake	970	Frontenac	K. & P. R.
Buda		Thunder Bay	
Station (summit)	1,472		C. P. R.
Oskondiga river, 0·9 mile, W., water, 1,421; rail	1,453		C. P. R.
Budd	619	Simcoe	G. T. R.
Buller	585	Hastings	C. P. R.
Burford	845	Brant	G. T. R.
Burgessville	913	Oxford	G. T. R.
Burketon	1,057	Durham	C. P. R.
Burks Falls		Parry Sound	
Station	974		G. T. R.
Summit, 4·7 miles, N., ground, 1,160; rail	1,139		G. T. R.
Burlington junction	328	Halton	G. T. R.
Burlington station	281	Halton	G. T. R.
Burnley village	653	Northumberland	C. N. & R.
Burnt lake	1,254	Nipissing	Geol. Su.
Burnt Island lake	1,340	Nipissing	Geol. Surv.
Burnt River (formerly Rettie)		Victoria	
Station	868		G. T. R.
Burnt river, 2·8 miles, S., water, 840; rail	861		G. T. R.
Burnt summit	980	Algoma	Admir. chart
Burritt	353	Grenville	C. P. R.
Burwash lake	934	Nipissing	Geol. Surv.
Bush lake	981	Nipissing	Geol. Surv.
Busteed	1,191	Rainy River	C. P. R.
Butler	1,423	Rainy River	C. P. R.
Buxton		Kent	
Mich. Cent. station	600		M. C. R.
L. E. & D. R. station	644		L. E. & D. R. R.
Cabot head	890	Bruce	Admir chart
Cache Bay station	652	Nipissing	C.P.R.
Cache lake	954	Thunder Bay	C.P.R.
Cache lake	1,406	Nipissing	C.A.R.
Cache Lake		Nipissing	
Station	1,417		C.A.R.
Madawaska river, 1·5 mile, E., water, 1,357; rail	1,383		C.A.R.
Summit, 2·8 miles, W., ground, 1509; rail	1,497		C.A.R.
Cache Lake station	968	Thunder Bay	C.P.R.
Caginagokog lake	1,461	Algoma	C.P.R.
Cainsville		Brant	
Grand Trunk station	720		G.T.R.
T. H. & B. station	707		T.H. & B.R.
Summit, 1·5 mile, N.	730		G.T.R.
Grand Trunk Ry. crossing	723		T.H. & B.R.
Calabogie		Renfrew	
Station	522		K. & P.R.
Summit, 1·2 mile, N.	578		K. & P.R.
Calabogie lake	507	Renfrew	K. & P.R.
Caldwell	546	Renfrew	C.A.R.
Caldwell lake	890	Lanark	K. & P.R.
Caledon	1,356	Peel	C.P.R.
Caledon East	946	Peel	G.T.R.

ONTARIO

Locality	Elev.	County	Authority
Caledonia..		Haldimand	
Station	652		G.T.R.
Grand river, water, 616 ; rail..............	652		G.T.R.
Summit, 2·8 miles, N.	681		G.T.R.
Caledonia Springs..	168	Prescott	C.P.R.
Callander.	671	Parry Sound	G.T.R.
Calvin.	696	Nipissing	C.P.R.
Cambray....	871	Victoria	G.T.R.
Cambridge.	228	Russell	N.Y. & O.R.
Camden East..	469	Lennox & Add'ton	B. of Q.R.
Cameron....	897	Victoria	G.T.R.
Cameron lake.	835	Victoria	Rys. & Canals
Cameron lake........	803	Nipissing	Geol. Surv.
Cameron or Round lake....	628	Lanark	K. & P.R.
Camlachie..	648	Lambton	G.T.R.
Campbellford		Northumberland	
Grand Trunk station.... .	496		G.T.R.
Proposed station......................	480		C.N. & P.R.
Trent river, high water, 466 ; low water, 461 ; rail..	492		G.T.R
Campbell lake....	648	Nipissing	Geol. Surv.
Campbell Cross	939	Peel	C.P.R.
Campbellville	924	Halton	C.P.R.
Canfield .		Haldimand	
Grand Trunk station. ...	623		G.T.R.
Mich. Cent. station...........	616		M.C.R.
Canfield junction......	616	Haldimand	G.T.R.
Cannington ...	850	Ontario	G.T.R.
Canoe lake.	465	Frontenac	K. & P.R.
Canoe lake. ..	1,392	Nipissing	C.A.R.
Canoe Lake station.........	1,379	Nipissing	C.A.R.
Canyon.	608	Renfrew	K. & P.R.
Cape Rich bluff.	1,005	Grey	Admir. chart
Caradoc.	804	Middlesex	C.P.R.
Cardinal....		Grenville	
Station	276		G.T.R.
B.M., west wall of culvert, quarter mile. W. of station	273		D. W. Com.
B.M., north coping of culvert, quarter mile, E. of station.......	269		D. W. Com.
Cardwell..		Peel	
Station	950		C.P.R.
Grand Trunk crossing, G.T.R., rail, 948; C.P.R., rail	968		C.P.R.
Summit, 0·7 mile, N.	965		G.T.R.
Cargill.	874	Bruce	G.T.R.
Caribou lake (Maniton lake)...........	1,236	Rainy River	C.P.R.
Caribou lake (Summit station).	1,326	Rainy River	C.P.R.
Carleton Place.		Lanark	
Station	447		C.P.R.
Mississippi river, high water, 433 ; low water, 427 ; rail	447		C.P.R.
Carlow road..	1,073	Hastings	C.O.R.
Carlstadt .	1,515	Thunder Bay	C.P.R.
Carlton.	405	York	G.T.R.
Carmel .	906	Durham	G.T.R.
Carp	311	Carleton	C.A.R.
Carp lake ...	1,329	Rainy River	Dawson
Carry.	1,226	Algoma	C.P.R.
Carson lake.	926	Renfrew	C.A.R.
Carthew..	814	Simcoe	G.T.R.
Cartier .	1,381	Algoma	C.P.R.
Casselman.		Russell	
Station	208		C.A.R.
South Nation river, high water, 188 ; low water, 181 ; rail	208		C.A.R.
Castleford.	307	Renfrew	C.P.R.
Caswell lake.	1,001	Muskoka	G.T.R.
Cat lake.	1,360	Rainy River	Dawson
Cat lake.	1,372	Algoma	C.P.R.
Cataract .	1,252	Peel	C.P.R.

ONTARIO

Locality	Elev.	County	Authority
Caterpillar lake	1,350	Rainy River	C.P.R.
Catfish lake	1,224	Nipissing	Geol. Surv.
Cargville	645	Durham	C.P.R.
Cape point	720	Bruce	Admir. chart
Cayuga		Haldimand	
Grand Trunk station	610		G.T.R.
Mich. Cent. station	625		M.C.R.
Grand river, high water, 585 ; low water, 578 ; rail	614		G.T.R.
Grand river, bed, 574 ; rail	624		M.C.R.
Summit, 2·5 miles, E.	660		M.C.R.
Cedar lake	984	Nipissing	Geol. Surv.
Cedar Springs	660	Kent	L.E. & D.R.R.
Cement Mill	451	Welland	M.C.R.
Centralia	867	Huron	G.T.R.
Central Ontario junction		Hastings	
Station	508		C.P.R.
Crow river, 3·4 miles, W., water, 502 ; rail	508		C.P.R.
Centreville	927	Peel	G.T.R.
Chalk River		Renfrew	
Station	524		C.P.R.
Chalk river, 2·8 miles, S., water, 477 ; rail	485		C.P.R.
Chamberlain	638	Lambton	M.C.R.
Chantler	603	Welland	T.H. & B.R.
Chapleau		Algoma	
Station	1,418		C.P.R.
Kebsquashesing lake, high water, 1,407 ; low water	1,403		C.P.R.
Charing Cross	625		M.C.R.
Charleston lake	272	Kent	Rys. & Canals
Chatauqua	289	Leeds	M.C.R.
Chatham		Lincoln	
		Kent	
Grand Trunk station	598		G.T.R.
Can. Pac. station	592		C.P.R.
L. E. & D. R. station	595		L. E. & D.R.R.
Chatham junction	577		G.T.R.
Thames river, high water, 589 ; low water, 575 ; rail	577		L.E. & D.R.R.
Sixth St. bridge	576		City Eng.
Fifth St. bridge	600		City Eng.
Aberdeen bridge	600		City Eng.
Waterworks standpipe, top	703		City Eng.
Sedimentation basin, top of embankment	604		City Eng.
City hall, ground floor	608		City Eng.
Post office, plinth	599		City Eng.
Chats lake, high water, 246 ; low water	239	Renfrew	Rys. & Canals
Chatsworth	944	Grey	C.P.R.
Chaudière junction	206	Carleton	C.A.R.
Chaudière junction	274	Carleton	C.P.R.
Chebogamog lake	798	Nipissing	Geol. Surv.
Chelmsford	884	Algoma	C.P.R.
Cheltenham		Peel	
Can. Pac. station	936		C.P.R.
Grand Trunk station	922		G.S.R.
Chemong station (disused)	816	Peterborough	C.T.R.
Chemong lake	805	Peterborough	G.T.R.
Cheney	211	Russell	C.A.R
Chesley		York	
Station	980		G.T.R.
Summit, 0·9 mile, S., ground, 1,022 ; rail	1,002		C.T.R.
Chesterville		Dundas	
Station	238		C.P.R.
South Nation river, 1·2 mile, E., bed, 200 ; rail	238		C.P.R.
Chin, Cape	727	Bruce	Admir.
Chippawa	572	Welland	M.C.R.
Chisholm		Hastings	
Station	389		C.O.R.
river, water, 397 ; rail	384		C.O.R.
Christian Island	880	Simcoe	Admir. chart
Churches		Brant	
Station	829		T.H. & B.R.
Summit at gravel pits 0·5 mile, N.	845		T.H. & B.R.
Church hill	880	Algoma	Geol. Surv.

ONTARIO

Locality	Elev.	County	Authority
Churchville	604	Peel	C.P.R.
Clandeboye	853	Middlesex	G.T.R.
Clarabelle junction	919	Nipissing	M. & N. S. R.
Clarabelle lake	956	Nipissing	C. C. Co.
Clare lake	1,284	Rainy River	C.P.R.
Claremont (summit)	396	Ontario	C.P.R.
Clarence Creek station	271	Russell	C.A.R.
Clarendon		Frontenac	
Station	764		K. & P. R.
Summit 0.6 mile, N	781		K. & P. R.
Clark	415	Leeds	C.P.R.
Clarkson		Peel	G.T.R.
Clay Banks bluff		Grey	Admir. chart
Clear lake			Rys. & Canals
Clear lake	708	Peterborough	Rys. & Canals
Clear lake, high water, 750 ; low water	755	Renfrew	K. & P. R.
Clearwater lake	1,357	Rainy River	Dawson
Cliff lake	1,048	Nipissing	Geol. Surv.
Clifford	1,226	Wellington	G.T.R.
Clifton		Welland	
Mich. Cent. station	582		M.C.R.
Grand Trunk station	582		G.T.R.
Mich. Cent. bridge, rail	586		M.C.R.
Clinton station	912	Huron	G.T.R.
Clinton junction	905	Huron	G.T.R.
Cloche bluff		Algoma	Admir. chart
Clontarf P.O.	865	Renfrew	K. & P. R.
Club lake, high water, 1256 ; low water	1,250	Rainy River	C.N.R.
Clyde Forks		Lanark	
Station	648		K. & P. R.
Clyde river, water	606		K. & P. R.
Clyde lake	642	Renfrew	K. & P. R.
Coatsworth	629	Kent	L. E. & D. R. R.
Cobden	476	Renfrew	C.P.R.
Coboconk	855	Victoria	G.T.R.
Cobourg	295	Northumberland	G.T.R.
Cochill	1,050	Hastings	C.O.R.
Coffin hill	954	Grey	Admir. chart
Colborne	321	Northumberland	G.T.R.
Coldwater	688	Simcoe	G.T.R.
Coldwater lake	1,396	Thunder Bay	Dawson
Coldwell	790	Thunder Bay	C.P.R.
Cole lake	1,044	Nipissing	Geol. Surv.
Collingwood		Simcoe	
Grand Trunk station	587		G.T.R.
H. & N. W. station (old)	592		G.T.R.
Collins Bay station	284	Frontenac	G.T.R.
Colwell	747	Simcoe	G.T.R.
Comber	606	Essex	M.C.R.
Commanda lake	738	Parry Sound	C.P.R.
Como, Lake	1,501	Algoma	C.P.R.
Concon	255	Prince Edward	C.O.R.
Conecon lake	263	Prince Edward	C.O.R.
Conger hill	1,436	Renfrew	K. & P. R.
Consecon lake	637	Renfrew	K. & P. R.
Conessee	955	Haliburton	I. B. & O. R.
Cook	642	Haldimand	G.T.R.
Cookstown	759	Simcoe	G.T.R.
Cooksville	391	Peel	C.P.R.
Cooper lake	914	Nipissing	Geol. Surv.
Copetown		Wentworth	
Station	755		G.T.R.
Summit of deep hill, 0.9 mile, E., ground, 773; rail	735		G.T.R.
Copper Cliff		Nipissing	
New station	849		C.P.R.
Old station	843		C.P.R.
Evans mine, foundation of engine house	887		C.C. Co.
Top of smelter stack	897		C.C. Co.
Door-sill of machine shop	850		C.C. Co.
Matte track, opposite coke shed	852		C.C. Co.

ONTARIO

Locality	Elev.	County	Authority
Corbetton	1,661	Dufferin	C. P. R.
Corbyville..........................	333	Hastings	G. T. R.
Corinth.............................	768	Elgin	G. T. R.
Cornell	795	Oxford	M. C. R.
Cornwall...........................		Stormont	
Grand Trunk station................	191		G. T. R.
N. Y. & O. .tion	213		N. Y. & O. R.
B. M., west wall of culvert at 67·1 miles ...			D. W. Con.
B. M., on station			D. W. Con.
New lower entrance lock, lower sill 138·20; coping.	189·4		D. W. Con.
Old lower entrance lock, lower sill	195·6		D. W. Con.
River St. Lawrence—			
Standard high water (1870)	155·0		D. W. Con.
Extreme low water	151·7		D. W. Con.
Extreme low water (Nov, 21, 1895)	150·5		D. W. Con.
Cornwall junction	201	Stormont	G. T. R.
Corson siding.......................	880	Victoria	G. T. R.
Corunna............................	629	Lambton	L. E. & D. R. R.
Corwhin	1,127	Wellington	C. P. R.
Couchiching lake, flood (May, 1871) 728; low water	719	Simcoe	G. T. R.
Couchie lake........................	1,142	Parry Sound	C. P. R.
Courtland...........................	774	Norfolk	G. T. R.
Courtright		Lambton	
Mich. Cent. station	588		M. C. R.
L. E. & D. R. station...............	601		L. E. & D. R. R.
Courtright junction..................	610	Lambton	M. C. R.
Crab or Crook lake...................	1,348	Algoma	Geol. Surv.
Craigleith	508	Grey	G. T. R.
Craigvale...........................	880	Simcoe	G. T. R.
Cranberry lake	323	Frontenac	Rys. & Canals
Crayfish	1,539	Thunder Bay	C. N. R.
Crayfish lake........................	1,535	Thunder Bay	C. N. R.
Creemore	850	Simcoe	G. T. R.
Creighton mine, foundation of rock-house.	973	Algoma	C. C. Co.
Cresswell	977	Victoria	G. T. R.
Croix, Lac la or Nequaquon lake	1,181	Rainy River	Dawson
Crombie............................	1,600	Dufferin	C. P. R.
Crooked lake........................	953	Nipissing	Geol. Surv.
Crooked lake........................	1,241	Rainy River	Dawson
Crookston...........................	563	Hastings	G. T. R.
Crosby	405	Leeds	B. & W. R.
Cross lake	959	Nipissing	Geol. Surv.
Cross lake...........................	646	Lennox & Add'ton	Geol. Surv.
Crossing lake........................	1,439	Rainy River	C. N. R.
Crow bay	528	Northumberland	Rys. & Canals
Crow lake...........................	535	Frontenac	K. & P. R.
Crumlin	890	Middlesex	C. P. R.
Crysler		Stormont	
Station	222		N. Y. & O. R.
South Nation river, water, 193; rail........	222		N. Y. & O. R.
Currie..............................		Oxford	
Station	985		G. T. R.
Summit, 0·3 mile, S................	1,040		G. T. R.
Cypress lake.........................	1,385	Rainy River	Dawson
Cypress or Wagong lake..............	1,428	Algoma	C. P. R.
Dagmar.............................	858	Ontario	C. P. R.
Dalkeith............................	232	Glengarry	C. A. R.
Dalton..............................	1,162	Algoma	C. P. R.
Darling Road........................	603	Haldimand	G. T. R.
Darlington...........................	379	Durham	G. T. R.
Darrell	600	Kent	L. E. & D. R. R.
Davenport...........................	422	York	G. T. R.
Dayton.............................	607	Algoma	C. P. R.
Dean Lake station	629	Algoma	C. P. R.
Deception	1,216	Rainy River	C. P. R.
Deception lake	1,094	Rainy River	C. P. R.
Deecwsville	685	Haldimand	G. T. R.
Deer bay	793	Peterborough	Rys. & Canals

ONTARIO

Locality	Elev.	County	Authority
Deer lake	822	Parry Sound	C. P. R.
Deer Lake station	1,233	Haliburton	I. B. & O. R.
Delhi	761	Norfolk	G. T. R
Delmage lake	1,385	Algoma	C. P. R.
Delta	312	Leeds	B. & W. R.
Delta lake	295	Leeds	Rys. & Canals
Denedus lake	1,022	Nipissing	Geol. Surv.
Denfield	897	Middlesex	G. T. R.
Denison	1,174	Thunder Bay	C. P. R.
Depot Harbour	590	Parry Sound	C. A. R.
Desbarats	595	Algoma	C. P. R.
Deschenes lake, high water, 198 ; low water	189	Carleton	Rys. & Canals
Deseronto station	252	Hastings	B. of Q R.
Deseronto junction	325	Lennox & Add'ton	G. T. R.
Desert lake	450	Frontenac	K. & P. R.
Desperation lake	983	Nipissing	Geol. Surv.
Deux Rivieres		Renfrew	
Station	518		C. P. R.
Ottawa river (Kelly bay) high water (May 10, 1880) 482 ; low water	469		C. P. R.
Devil lake, high water, 438 ; low water	431	Frontenac	Rys. & Canals
Dexter		Thunder Bay	
Station	1,582		C. P. R.
Summit, 0·7 mile, N.	1,585		C. P. R.
Diamond lake	892	Parry Sound	C. A. R.
Diamond lake	1,215	Hastings	I. B. & O. R.
Diltz	586	Haldimand	G. T. R.
Dinorwic	1,239	Rainy River	C. P. R.
Dinorwic or Little Wabigoon lake	1,206	Rainy River	C. P. R.
Dixie		Peel	
Station	375		C. P. R.
Etobicoke river, 0·9 mile, E., bed, 333 ; rail	375		C. P. R.
Dodd hill	1,012	Grey	Admir. chart
Doe lake	968	Parry Sound	Geol. Surv.
Dog lake	323	Frontenac	Rys. & Canals
Dog lake	1,091	Algoma	C. P. R.
Dog or Sakwite lake	1,199	Rainy River	C. P. R.
Dog lake	1,331	Thunder Bay	C. P. R.
Dog lake, Great	1,378	Thunder Bay	Dawson
Dog lake, Little	1,030	Thunder Bay	Dawson
Don	252	York	G. T. R.
Donald lake	1,473	Thunder Bay	C. P. R.
Doon	953	Waterloo	G. T. R.
Dorchester	852	Middlesex	G. T. R.
Doré lake, low water	468	Renfrew	K. & P. R.
Doré lake	1,338	Rainy River	Dawson
Douglas		Renfrew	
Can. Pac. station	417		C. P. R.
Can. Atl. station	493		C. A. R.
Bonnechere river, water, 405 ; rail	422		C. P. R.
Downsview	573	York	G. T. R.
Drayton		Wellington	
Station	1,356		G. T. R.
Conestoga creek, water, 1,300 ; rail	1,350		G. T. R.
Dresden	604	Kent	L. E. & D. R. R.
Drumbo		Oxford	
Can. Pac. station	1,011		C. P. R.
Grand Trunk station	1,009		G. T. R.
Grand Trunk Ry. crossing	1,011		C. P. R.
Drummond	694	Peterborough	G. T. R.
Dryden		Rainy River	
Station	1,220		C. P. R.
Wabigoon river, water, 1,182 ; rail	1,220		C. P. R.
Dublin	1,103	Perth	G. T. R.
Duchesnay (now Roberts)	1,377	Algoma	C. P. R.
Duck lake	746	Parry Sound	C. A. R.
Duck lake	1,487	Thunder Bay	C. P. R.
Dufferin	684	Haldimand	M. C. R.

ONTARIO

Locality	Elev.	County	Authority
Dumfries		Waterloo	
Can. Pac. station......	1.025		C. P. R.
Summit, 2·6 miles, E., rail......	1,049		C. P. R.
Dumfries	810	Brant	G. T. R.
Dunbarton.............	287	Ontario	G. T. R.
Duncan lake.............	971	Nipissing	Geol. Surv.
Duncanville village....	247	Russell	T. & O. R.
Dundalk........		Grey	
Station............	1,700		C. P. R.
Summit, 0·3 mile, N., ...	1,706		C. P. R.
Dundas....		Wentworth	
Grand Trunk station.. . . .	516		G. T. R.
T. H. & B. station.........	349		T. H. & B. R.
Spencer creek, bed, 406; rail....	507		G. T. R.
Dundas, Cape......	880	Bruce	Admir. chart
Dunkeld..........	954	Bruce	G. T. R.
Dunnville..........	584	Haldimand	G. T. R.
Duntroon............	938	Simcoe	G. T. R.
Dutchman head.......	760	Algoma	Admir. chart
Dutton.........		Elgin	
Mich. Cent. station....	721		M. C. R.
L. E. & D.R. station....	725		L. E. & D. R. R.
Dyment......................	712	Simcoe	G. T. R.
Dyment..	1,350	Rainy River	C. P. R.
Dynes.........	256	Wentworth	G. T. R.
Dysart.......		Haliburton	
Station..........	1,097		G. T. R.
Summit, 2·4 miles, N....	1,126		G. T. R.
Eagle lake....	631	Frontenac	K. & P. R.
Eagle lake....	1,164	Parry Sound	C. P. R.
Eagle lake.. . .	1,188	Rainy River	C. P. R.
Eagle River station...	1,186	Rainy River	C. P. R.
East bluff (Manitoulin island).....	834	Algoma	Admir. chart
East knob........	1,330	Algoma	Admir. chart
East lake....	351	Leeds	B. & W. R.
Eastman	225	Carleton	C. A. R.
Eastwood ...	973	Oxford	G. T. R.
Eau Claire.. .	591	Nipissing	C. P. R.
Eberts	600	Kent	L. E. & D. R. R.
Echo Bay station......	590	Algoma	C. P. R.
Echo lake....	1,434	Algoma	C. P. R.
Eden........		Elgin	
Station	783		T. L. E. & P. R
Summit, 1·6 mile, N........	794		T. L. E. & P. R.
Edgar........	626	Essex	M. C. R.
Edgemont....	544	Wentworth	T. H. & B. R.
Elgington..	891	Parry Sound	C. A. R.
Edwards....	258	Carleton	N. Y. & O. R.
Egan Estate	1,093	Nipissing	C. A. R.
Eganville		Renfrew	
Can. Pac. station...	549		C. P. R.
Can. Atl. station......	574		C. A. R.
Bonnechere river at road bridge, water, 531 ; bridge	546		Interior
Road opposite post office	554		Interior
Summit of road to Can. Atl. Ry. station	628		Interior
Egg lake....	1,083	Thunder Bay	Geol. Surv.
Ekfrid	709	Middlesex	G. T. R.
Elbe....	374	Leeds	B. & W. R.
Elbow lake....	574	Frontenac	K. & P. R.
Elbow lake. ...	678	Nipissing	Geol. Surv.
Elbow lake	1,435	Rainy River	C N. R.
Elder........	629	York	C. P. R.
Eldon ..	880	Victoria	G. T. R.
Eldorado........	788	Hastings	C O R.
Elgin....	432	Leeds	B. & W. R.
Elia....	673	York	G. T. R.
Elm Bay spur	1,239	Rainy River	C. P. R.

ONTARIO

Locality	Elev.	County	Authority
Elmira	1,142	Waterloo	G.T.R.
Elmsley	438	Lanark	C.P.R.
Elmstead	604	Essex	C.P.R.
Elmvale	608	Simcoe	G.T.R.
Elora		Wellington	
Can. Pac. station	1,270		C.P.R.
Grand Trunk station	1,292		G.T.R.
Elsinore	257	Wentworth	G.T.R.
Elsie junction	960	Nipissing	M. & N.S.R.
Embro		Oxford	
Station	1,004		C.P.R.
Summit, 2·5 miles, E.	1,017		C.P.R.
Embrun		Russell	
Station	223		N.Y. & O.R.
Castor brook, water, 192; rail.	222		N.Y. & O.R.
Emerald lake	1,009	Nipissing	Geol. Surv.
Emery	520	York	C.P.R.
Emmett	600	Kent	L.E. & D.R.R.
Emo	1,097	Rainy River	C.N.R.
Emsdale	1,042	Parry Sound	G.T.R.
English River		Rainy River	
Station	1,518		C.P.R.
English river, water, 1,511; rail	1,516		C.P.R.
Enterprise	483	Lennox & Add'ton	B. of Q.R.
Erie, Lake			
Standard high water (extreme high water, June 1838).	574·9		D.W. Com.
Extreme low water (1810 and 1811)	568·9		D.W. Com.
Standard low water	571·0		D.W. Com.
Mean water, 1860 to 1900.	572·4		D.W. Com.
Mean water, 1871 to 1900.	572·3		D.W. Com.
Erin	1,291	Wellington	C.P.R.
Erindale	474	Peel	C.P.R.
Erinsville		Lennox & Add'ton	
Station	549		B. of Q. R.
Summit, 1·2 mile, E.	587		B. of Q. R.
Ernestown	324	Lennox & Add'ton	G.T.R.
Esher		Algoma	
Station	1,489		C.P.R.
Summit, 3·1 miles, W., ground, 1500; rail	1,541		C.P.R.
Esnagami or Shell lake	1,273	Algoma	C.P.R.
Essex	643	Essex	M.C.R.
Esther cliff	952	Grey	Admir. chart
Ethel		Huron	
Station	1,174		G.T.R.
Middle branch Maitland river, bed, 1135; rail	1,166		G.T.R.
Ettrick	901	Middlesex	G.T.R.
Eureka Lake station	1,391	Algoma	C.P.R.
Eureka lake, high water, 1384; low water.	1,376	Algoma	C.P.R.
Evans mine, foundation of engine house	847	Nipissing	C. C. Co.
Everett	795	Simcoe	G.T.R.
Exeter		Huron	
Station	872		G.T.R.
River Ausable, 0·5 mile, N., bed, 831; rail	864		G.T.R.
Expectation lake	968	Nipissing	Geol. Surv.
Fairbank	511	York	G. T. R.
Fairbank or Washaigamag lake.	867	Algoma	Geol. Surv.
Fairfield	395	Leeds	C. P. R.
Fairfield	899	Middlesex	G. T. R.
Fairy lake	936	Muskoka	G. T. R.
Falcon	1,509	Rainy River	C. P. R.
Falding	770	Parry Sound	C. A. R.
Falkenburg		Muskoka	
Station	974		G. T. R.
Summit, 1·7 mile, N., ground, 1069; rail	1,058		G. T. R.
Falls View	616	Welland	M. C. R.
Fanny lake	984	Nipissing	Geol. Surv.
Fargo	637	Kent	M. C. R.
Farmersville (now Athens).	416	Leeds	B. & W. R.

ONTARIO

Locality	Elev.	County	Authority
Farran Point.		Stormont	
Station	241		G. T. R.
B.M., south coping of culvert ½ mile, W. of station	242·0		D. W. Com.
New lock, lower sill	181·0		D. W. Com.
Old lock, lower sill, 192·69 ; coping	209·2		D. W. Com.
Standard high water (1870)	202·3		D. W. Com.
Extreme low water (Nov. 1895).	197·3		D. W. Com.
Farrington	1,357	Rainy River	C. N. R.
Feist lake	1,328	Rainy River	C. P. R.
Fell	854	Victoria	G. T. R.
Fenelon Falls		Victoria	
Station	846		G. T. R.
Trent Valley canal, water, 837 ; rail.	851		G. T. R.
Fenwick	620	Welland	T. H. & B. R.
Fergus		Wellington	
Can. Pac. station	1,355		C. P. R.
Grand Trunk station	1,356		G. T. R.
Can. Pac. Ry. crossing	1,363		G. T. R.
Grand river, 0·9 mile, W., high water, 1,250 ; low water, 1,342 ; rail.	1,301		G. T. R.
Ferguson		Middlesex	
Station	697		G. T. R.
Bear creek, 1·0 mile, N., high water, 640 ; low water, 635 ; rail	690		G. T. R.
Ferguson lake	971	Nipissing	Geol. Surv.
Fesserton	589	Simcoe	G. T. R.
Field	610	Grey	G. T. R.
Finch		Stormont	
Station	273		C. P. R.
Payne river, bed, 254 ; rail	272		C. P. R.
Finmark		Thunder Bay	
Station	1,180		C. P. R.
Sunshine brook, 1·5 mile, W., low water, 1,158 ; high water 1,162 ; rail	1,168		C. P. R.
Fire Hill station	606	Thunder Bay	C. P. R.
Fleming lake	1,084	Thunder Bay	Geol. Surv.
Flesherton	1,557	Grey	C. P. R.
Fletcher	597	Kent	M. C. R.
Flower.		Lanark	
Station	626		K. & P. R.
Summit, 3·7 miles, N	654		K. & P. R.
Folger	810	Lanark	K. & P. R.
Fordwich	1,194	Huron	C. P. R.
Forest	711	Lambton	G. T. R.
Forestville	710	Norfolk	G. T. R.
Forks (formerly Spanish Forks).	1,203	Algoma	C. P. R.
Forks lake	1,548	Thunder Bay	C. N. R.
Forks of Credit		Peel	
Station	1,076		C. P. R.
Credit river, bed, 1,005 ; rail	1,076		C. P. R.
Fort Erie		Welland	
South yard	605		G. T. R.
Grand Trunk station	605		G. T. R.
Mich. Cent. station	595		M. C. R.
Fort Frances	1,112	Rainy River	C. N. R.
Forthton (summit)	409	Leeds	B. & W. R.
Fort William		Thunder Bay	
Can. Nor. station	612		C. N. R.
Can. Pac. station	607		C. P. R.
Fort William West	635	Thunder Bay	C. P. R.
Fourth Chute	437	Renfrew	C. P. R.
Fowl lake	1,438	Thunder Bay	Dawson
Foxboro		Hastings	
Station	348		G. T. R.
Moira river, water, 332 ; rail	347		G. T. R.
Foxmead	677	Simcoe	G. T. R.
Frankford		Hastings	
Station	371		C. O. R.
Trent river, water.	361		Rys. & Canals
Franklin	893	Durham	G. T. R.

ONTARIO

Locality	Elev.	County	Authority
Franktown station	476	Lanark	C. P. R.
Franktown village	482	Lanark	T. & O. R.
Fraserville		Peterborough	
Station	638		G. T. R.
Summit, 1·3 mile, S	680		G. T. R.
Frazer Bay hill	1,210	Algoma	Admir. chart
Fredericksburg	307	Lennox & Add'ton	G. T. R.
French lake	1,346	Rainy River	Dawson
Friday lake	1,103	Nipissing	Geol. Surv.
Frontenac		Lennox & Add'ton	
Station	528		B. of Q. R.
Summit, 0·3 mile, E	531		B. of Q. R.
Fulton	1,229	Wellington	G. T. R.
Furnace Falls	970	Haliburton	I. B. & O. R.
Gage	279	Wentworth	G. T. R.
Galbraith	501	Lennox & Add'ton	B. of Q. R.
Galetta		Carleton	
Station	295		C. A. R.
Mississippi river, water, 272; rail	292		C. A. R.
Summit, 2·1 miles, S., ground, 337; rail	330		C. A. R.
Galt		Waterloo	
W. G. & B. station	875		G. T. R.
G. & E. station	878		G. T. R.
Can. Pac. station	931		C. P. R.
Grand Trunk Ry., crossing, G. T. Ry., 898; Can. Pac. Ry	927		C. P. R.
Grand river, bed, 856; rail	927		C. P. R.
Gamebridge	801	Ontario	G. T. R.
Gananoque		Leeds	
Grand Trunk station	306		G. T. R.
T. I. station	251		T. I. R.
Gananoque river, water, 266; rail	290		G. T. R.
B. M., on plinth course of station	309·3		D. W. Com.
B. M., on coping of west abutment of G. T. Ry. bridge over Gananoque river	291·2		D. W. Com.
Front door-sill of custom-house	256·0		D. W. Com.
Gananoque River station	265	Leeds	T. I. R.
Garden Hill station	626	Durham	G. T. R.
Garden River station	606	Algoma	C. P. R.
Gargantua hill	1,031	Algoma	Admir. chart
Garnett	698	Haldimand	G. T. R.
Garson lake, low water	926	Nipissing	
Garwood	1,156	Rainy River	C. P. R.
Gelert	1,032	Haliburton	G. T. R.
Geneva		Algoma	
Station	1,360		C. P. R.
Summit, 1·0 mile, N., ground, 1,419; rail	1,408		C. P. R.
Geneva lake	1,345	Algoma	Geol. Surv.
Georgetown		Halton	
Station	846		G. T. R.
Credit river, 1·0 mile, E., water, 713; rail	827		G. T. R.
Georgetown junction	872	Halton	G. T. R.
German Mills	1,003	Waterloo	G. T. R.
Giants Tomb island	830	Simcoe	Admir. chart
Gibson	782	Norfolk	G. T. R.
Gibson Landing	254	Halton	G. T. R.
Gilbert	1,216	Rainy River	C. P. R.
Gilford	752	Simcoe	G. T. R.
Gilmour		Hastings	
Station	1,018		C. O. R.
Summit, 3·0 miles, S	1,035		C. O. R.
Girdwood (summit)	1,441	Algoma	C. P. R.
Glanford		Wentworth	
Station	724		G. T. R.
Summit, 0·6 mile, N	730		G. T. R.
Glanworth		Middlesex	
Station	872		L. E. & D. R. R.
Summit, 2·0 miles, N	918		L. E. & D. R. R.
Glasford lake	927	Nipissing	Geol. Surv.

ONTARIO

Locality	Elev.	County	Authority
Glasgow	445	Renfrew	C. A. R.
Glenannan	1,073	Huron	C. P. R.
Glencairn	738	Simcoe	G. T. R.
Glencoe	728	Middlesex	G. T. R.
Glendower mines	520	Frontenac	K. & P. R.
Glen Huron		Simcoe	
Station	1,034		G. T. R.
Summit, 0.5 mile, W., ground, 1,056; rail	1,052		G. T. R.
Glen Major		Ontario	
Station	845		C. P. R.
Duffin Brook, bed, 780; rail	830		C. P. R.
Glen Norman	251	Glengarry	C. P. R.
Glenorchy		Rainy River	
Station	1,172		C. N. R.
Summit, 3.7 miles, E., ground, 1,232; rail	1,223		C. N. R.
Glen Rae	681	Lambton	M. C. R.
Glen Robertson	261	Glengarry	C. A. R.
Glenroy	267	Glengarry	C. P. R.
Glen Sandfield	240	Glengarry	C. A. R.
Glenvale		Frontenac	
Station	420		K. & P. R.
Collins brook, 2.9 miles, S., water, 284; rail	302		K. & P. R.
Glen Williams	913	Halton	G. T. R.
Glenwood	628	Kent	L. E. & D. R. R.
Gloucester		Carleton	
Station	345		C. P. R.
Summit, 0.2 mile, S., ground, 367; rail	353		C. P. R.
Goble	932	Oxford	G. T. R.
Goderich	729	Huron	G. T. R.
Golden lake	560	Renfrew	C. A. R.
Golden Lake junction	595	Renfrew	C. A. R.
Goldstone		Wellington	
Station	1,456		G. T. R.
Summit, 2.3 miles, E.	1,524		G. T. R.
Gooderham		Haliburton	
Station	1,073		I. B. & O. R.
Burnt river, 0.3 mile, E., bed, 1,060; rail	1,074		I. B. & O. R.
Goodwood		Ontario	
Station	1,102		G. T. R.
Summit, 0.8 mile, N.	1,152		G. T. R.
Gordon	581	Essex	M. C. R.
Gordon Lake		Algoma	
Station	646		B. M. & A. R.
Summit, 1.1 mile, N.	687		B. M. & A. R.
Gordon lake	637	Algoma	B. M. & A. R.
Gorrie		Huron	
Station	1,130		C. P. R.
Maitland river, 1.1 mile, E., bed, 1,110; rail	1,127		C. P. R.
Goshen	497	Renfrew	C. A. R.
Gould lake	458	Frontenac	K. & P. R.
Gould	1,081	Haliburton	G. T. R.
Gourock	1,059	Wellington	G. T. R.
Government Road		Renfrew	
Station	414		C. P. R.
Summit, 1.9 mile, S.	467		C. P. R.
Governors Road	967	Oxford	G. T. R.
Gowan		Simcoe	
Station	819		G. T. R.
Summit, 1.2 mile, N., ground, 842; rail	839		G. T. R.
Gowanstown	1,282	Perth	G. T. R.
Grace lake	1,459	Algoma	C. P. R.
Grafton	283	Northumberland	G. T. R.
Graham		Renfrew	
Station	420		C. P. R.
Muskrat river, water, 409; rail	417		C. P. R.
Graham lake	836	Lanark	K. & P. R.
Grand Valley		Dufferin	
Station	1,542		C. P. R.
Summit, 3.3 miles, W.	1,589		C. P. R.
Granite lake	1,006	Nipissing	Geol. Surv.

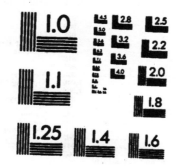

1653 East Main Street
Rochester, New York   14609   USA
(716) 482 - 0300 - Phone
(716) 288 - 5989 - Fax

ONTARIO

Locality	Elev.	County	Authority
Granite mountain....	1,300	Algoma	Geol. Surv.
Granton....	1,033	Middlesex	G. T. R.
Grassett....	1,246	Thunder Bay	C. P. R.
Grass Hill		Victoria	
Station	949		G. T. R.
Summit, 0·8 mile, W	968		G. T. R.
Grassie (summit)....	651	Lincoln	T. H. & B. R.
Gravel....	628	Thunder Bay	C. P. R.
Gravenhurst		Muskoka	
Station....	818		G. T. R.
Summit, 2·0 miles, S.	851		G. T. R.
Greenfield....	343	Glengarry	C. A. R.
Green lake....	1,046	Nipissing	Geol. Surv.
Green lake....	1,362	Algoma	C. P. R.
Green Valley....	279	Glengarry	C. P. R.
Griffith island....	880	Bruce	Admir. chart
Grimsby....	287	Lincoln	G. T. R.
Grimsby Park....	302	Lincoln	G. T. R.
Guelph....		Wellington	
Grand Trunk station....	1,067		G. T. R.
Can. Pac. station....	1,041		C. P. R.
City hall, floor level....	1,063		City Eng.
Highest point, near corner Grange and Tryeathlen streets....	1,143		City Eng.
Waterworks, ordinary low water in Eramosa river, at intake	1,011		City Eng.
Speed river			
At N.W. boundary of city, ordinary water	1,054		City Eng.
At Grand Trunk bridge (rail, 1,066) water	1,028		G. T. R.
At junction of Eramosa river	1,010		City Eng.
At south corner of city	994		City Eng.
Guelph junction....	1,075	Wellington	G. T. R.
Guelph junction	960	Halton	C. P. R.
Gulch hill....	1,535	Algoma	Admir. chart
Gull lake....	700	Parry Sound	C. P. R.
Gull lake....	856	Frontenac	K. & P. R.
Gun lake....	1,034	Rainy River	Dawson
Gun lake	1,166	Nipissing	C. A. R.
Gunflint....		Thunder Bay	
Station....	1,569		C. N. R.
Summit, 6·2 miles, E	1,627		C. N. R.
Gunflint lake....	1,545	Thunder Bay	C. N. R.
Gurney (formerly Cypress)..	626	Algoma	C. P. R.
Hadley ..	1,120	Haliburton	I. B. & O. R.
Hagar	675	Nipissing	C. P. R.
Hagersville		Haldimand	
Grand Trunk station....	729		G. T. R.
Mich. Cent. crossing.	727		G. T. R.
Haley....		Renfrew	
Station....	530		C. P. R.
Summit, 1·4 mile, S....	543		C. P. R.
Haliburton....	1,050	Haliburton	G. T. R.
Haliburton Road....		Haliburton	
Station....	1,181		I. B. & O. R.
Summit, 0·9 mile, E., ground, 1,217; rail....	1,211		I. B. & O. R.
Halliday hill....	951	Grey	Admir. chart
Halloway	466	Hastings	G. T. R.
Hallowell....	253	Prince Edward	C. O. R.
Hall....	855	Victoria	G. T. R.
Hamburg		Waterloo	
Station	1,126		G. T. R.
River Nith, water, 1,083; rail.	1,124		G. T. R.
Hamilton		Wentworth	
Stuart street....	254		G. T. R.
King street....	306		G. T. R.
Wentworth street....	267		G. T. R.
T. H. & B. crossing	326		G. T. R.
T. H. & B. station....	325		T. H. & B. R.
Garth street	305		T. H. & B. R.

ONTARIO

Locality	Elev.	County	Authority
Hamilton.		Wentworth	
Poulette street junction	298		T. H. & B. R.
Wentworth street	300		T. H. & B. R.
Hunter street tunnel, ground at centre, 363 ; rail	319		T. H. & B. R.
Masonic hall, west side of centre door	312		City Eng.
Barton reservoir, water	433		City Eng.
Ferguson Ave. reservoir, water	543		City Eng.
City hall, floor level	323		City Eng.
Corner, Main & Hess streets (foot of mountain)	364		City Eng.
Mountain ridge	628		City Eng.
Desjardins canal, rail	288		G. T. R.
Junction Cut	318		G. T. R.
Hammond	220	Russell	C. A. R.
Hanging-stone lake	919	Nipissing	Geol. Surv.
Hargrove	258	Halton	G. T. R.
Harley	836	Brant	G. T. R.
Harrietsville.		Middlesex	
Station	949		C. P. R.
Summit, 0·4 mile, W	959		C. P. R.
Harrington	1,000	Bruce	G. T. R.
Harris, Lake	1,223	Rainy River	C. P. R.
Harrisburg	733	Brant	G. T. R.
Harrison	286	Stormont	N. Y. & O. R.
Harriston		Wellington	
Grand Trunk station	1,256		G. T. R.
Can. Pac. station	1,244		C. P. R.
Canadian Pacific Ry. crossing	1,248		G. T. R.
Maitland river, water, 1,244 ; rail	1,252		G. T. R.
Harrow	626	Essex	L. E. & D. R. R.
Harrowsmith		Frontenac	
Station	491		K. & P. R.
Summit, 1·0 mile, S	519		K. & P. R.
Harswell lake	585	Frontenac	K. & P. R.
Hartington	531	Frontenac	K. & P. R.
Hartney hill	780	Frontenac	K. & P. R.
Hastings		Peterborough	
Station	618		G. T. R.
Trent river, water, 609 ; rail	631		G. T. R.
Hatchley	850	Brant	G. T. R.
Havelock	700	Peterborough	C. P. R.
Hawk Lake station	1,290	Thunder Bay	C. P. R.
Hawk lake	1,580	Thunder Bay	C. P. R.
Hawkesbury	141	Prescott	C. A. R.
Hawkestone	781	Simcoe	G. T. R.
Hawthorne	220	Carleton	N. Y. & O. R.
Hawtrey.		Oxford	
Mich. Cent. station	782		M. C. R.
Grand Trunk station	786		G. T. R.
Summit, 1·6 mile, N	847		G. T. R.
Hay island	940	Bruce	Admir. chart
Hay lake		Algoma	
Standard high water	585·8		D. W. Com.
Standard low water	581·6		D. W. Com.
Mean water (1871 to 1900)	582·3		D. W. Com.
Hay lake	1,505	Thunder Bay	C. P. R.
Haycroft	593	Essex	C. P. R.
Hayes (summit)	359	Grenville	C. P. R.
Heidelberg	1,122	Waterloo	G. T. R.
Height of land, Ombabika river route (Hudson bay and St. Lawrence)	1,150	Thunder Bay	Geol. Surv.
Height-of-land lake	1,553	Thunder Bay	Dawson
Height-of-land, Kaministikwia route (Hudson bay and St. Lawrence)	1,586	Thunder Bay	Dawson
Height-of-land, Pigeon river route (Hudson bay and St. Lawrence)	1,595	Thunder Bay	Dawson
Helen, Lake	608	Thunder Bay	C. P. R.
Helen, Lake	1,138	Rainy River	C. P. R.
Helen mine	1,256	Algoma	A. C. & H. B. R.
Hematite	1,364	Rainy River	C. N. R.
Hendrie	734	Simcoe	G. T. R.

6

ONTARIO

Locality	Elev.	County	Authority
Henfryn...........	1,165	Huron	G. T. R.
Hensall...............	895	Huron	G. T. R.
Heron Bay.... , ...		Thunder Bay	
Station....	768		C. P. R.
Pic river, 1·1 mile, E., water, 603; rail...........	685		C. P. R.
Summit, 2·2 miles, W., ground, 778 · rail	752		C. P. R.
Heron lake	1,184	Rainy River	C. P. R.
Hespeler................		Waterloo	
Station ... .. .'	943		G. T. R.
Speed river, water, 935; rail......	948		G. T. R.
Hewitt	583	Welland	M. C. R.
Heywood island..........	758	Algoma	Admir. chart
Hickson (formerly Strathallan)....		Oxford	
Station ·	1,093		G. T. R.
Summit, 1·6 mile, N....·	1,131		G. T. R.
Highfield		York	G. T. R.
Highgate.... .		Kent	
Mich. Cent. station....	736		M. C. R.
L. E. & D. R. station....	740		L. E. & D. R. R.
High hill (Bidwell tp.).............	1,120	Algoma	Admir. chart
High Park	255	York	G. T. R.
High Point		Ontario	
Station	1,028		G. T. R.
Summit, 0·04 mile, S..........	1,029		G. T. R.
Hillcrest (closed)............	739	Nipissing	C. P. R.
Hillier		Prince Edward	
Station	296		C. O. R.
Summit, 0·6 mile, N....	314		C. O. R.
Hillsburg......	1,405	Wellington	C. P. R.
Hinchinbrooke...............	564	Frontenac	K. & P. R.
Hoard	381	Northumberland	G. T. R.
Hogan hill....	865	Algoma	Admir. chart
Holland Centre......		Grey	
Station	1,212		C. P. R.
Saugeen river, 0·5 mile, S., bed, 1,201; rail..	1,218		C. P. R.
Holland Landing.....	743	York	G. T. R.
Holliday hill............	626	Renfrew	K. & P. R.
Holmesdale.....	676	Lambton	M. C. R.
Holmesville	880	Huron	G. T. R.
Hope island..	870	Simcoe	Admir. chart
Hornby	649	Halton	C. P. R.
Horner.....	1,246	Rainy River	C. P. R.
Horsborough hill.........	770	Algoma	Admir. chart
Houghton	675	Elgin	T. L. E. & P. R.
Houghton lake.........	1,325	Algoma	C. P. R.
Howland	982	Haliburton	I. B. & O. R.
Humber.... .	264	York	G. T. R.
Humber Grove...	282	York	G. T. R.
Humberstone.....	591	Welland	G. T. R.
Humber Summit.....		York	
Station	498		C. P. R.
Humber river, bed, 441; rail..	498		C. P. R.
Hungerford....	554	Hastings	C. P. R.
Huntsville....		Muskoka	
Station.....	957		G. T. R.
Summit, 4·9 miles, S., ground, 1,066; rail......	1,049		G. T. R.
Huron, Lake....			
Standard high water(extreme high water, July, 1838)	584·4		D. W. Com.
Standard low water	579·7		D. W. Com.
Extreme low water (Feb, 1819)............	577·7		D. W. Com.
Mean water (1860 to 1900).....	581·1		D. W. Com.
Mean water (1871 to 1900) ...	581·0		D. W. Com.
Huronian (height-of-land).... ..	1,508	Thunder Bay	C. N. R.
Hyde Park....		Middlesex	
Can. Pac. station....	902		C. P. R.
Grand Trunk station ..	883		G. T. R.
Grand Trunk crossing, G. T. Ry., rail, 876; Can. Pac. Ry., rail....	902		C. P. R.
Hyland lake	1,160	Renfrew	K. & P. R.

ONTARIO

Locality	Elev.	County	Authority
Ignace		Rainy River	
Station	1,487		C. P. R.
Alginac river, 0·8 mile, E., bed, 1,467 ; rail	1,490		C. P. R.
Ilderton	934	Middlesex	G. T. R.
Isles, Lac des	1,568	Thunder Bay	C. P. R.
Indian lafle	403	Leeds	Rys. & Canals
Indian lake	901	Nipissing	Geol. Surv.
Indian River		Peterborough	
Station	709		C. P. R.
Indian river, bed, 661 ; rail	688		C. P. R.
Ingall lake	1,050	Nipissing	Geol. Surv.
Ingersoll		Oxford	
Can. Pac. station	879		C. P. R.
Grand Trunk station	879		G. T. R.
Summit, 2·0 miles, W	927		C. P. R.
Thames river, at Thames st. bridge, water	860		G. T. R.
Ingles	530	Welland	M. C. R.
Inglewood		Peel	
Station	896		G. T. R.
Credit river, 0·8 mile, N., water, 882 ; rail	893		G. T. R.
Ingoldsby (now Lochlin)	1,082	Haliburton	G. T. R.
Ingolf	1,185	Rainy River	C. P. R.
Innerkip		Oxford	
Station	969		C. P. R.
Thames river, bed, 947 ; rail	970		C. P. R.
Innisfil	819	Simcoe	G. T. R.
Inwood	688	Lambton	M. C. R.
Iona		Elgin	
Mich. Cent. station	745		M. C. R.
L. E. & D. R. station	745		L. E. & D. R. R.
Irish Creek station	342	Grenville	C. P. R.
Irondale station	980	Haliburton	I. B. & O. R.
Irondale junction	935	Haliburton	G. T. R.
Iron lake	1,215	Rainy River	Dawson
Iroquois		Dundas	
Station	242		G. T. R.
B. M., on station	246·1		D. W. Com.
B. M., north coping of culvert, quarter mile, W	242·0		D. W. Com.
Isbester	600	Algoma	C. P. R.
Island lake	681	Parry Sound	Geol. Surv.
Island lake	889	Nipissing	Geol. Surv.
Island lake	1,331	Rainy River	C. P. R.
Islington	400	York	C. P. R.
Ivanhoe		Hastings	
Station	608		C. P. R.
Grand Trunk Ry., crossing, G. T. Ry., rail, 589; Can. Pac. Ry., rail	584		C. P. R.
Jackfish	632	Thunder Bay	C. P. R.
Jackvine		Rainy River	
Station	1,376		C. P. R.
Summit, 2·0 miles, E	1,381		C. P. R.
Jackson Point	726	York	G. T. R.
James lake	1,023	Nipissing	Geol. Surv.
Janet head	728	Algoma	Admir. chart
Jarvis	699	Haldimand	G. T. R.
Jeanette Creek station	580	Kent	G. T. R.
Jelly	373	Leeds	C. P. R.
Jerseyville	738	Wentworth	T. H. & B. R.
Jigak lake	1,390	Rainy River	C. P. R.
Johnson	956	Lincoln	M. C. R.
Jones bluff	403	Bruce	Admir. chart
Jordan	309	Lincoln	G. T. R.
Joseph, Lake	748	Muskoka	Interior
Jumping Caribou lake	1,048	Nipissing	Geol. Surv.
Kachekatonga, Lake	1,386	Thunder Bay	C. P. R.
Kaibagou		Thunder Bay	
Station	1,486		C. N. R.
Summit, 0·9 mile, W	1,542		C. N. R.
Kaijikmaniton or Baptiste lake	1,205	Hastings	I. B. & O. R.

6½

ONTARIO

Locality	Elev.	County	Authority
Kakabeka (summit)	1,084	Thunder Bay	C. P. R.
Kakabeka Falls		Thunder Bay	
Station	915		C. N. R.
Kaministikwia river, water, 906 ; rail	918		C. N. R.
Kakakwipitchwi lake	905	Thunder Bay	C. P. R.
Kaladar		Lennox & Add'ton	
Station	702		C. P. R.
Summit, 0·3 mile, W., ground, 714 ; rail	706		C. P. R.
Kalmar	1,218	Rainy River	C. P. R.
Kaminiskaik lake	931	Renfrew	Interior
Kaministikwia		Thunder Bay	
Station	1,013		C. P. R.
Kaministikwia river, high water, 997 ; low water, 993 ; rail	1,013		C. P. R.
Kaogassikok or Pickerel lake	1,337	Rainy River	C. P. R.
Kaotisinimigouang lake	948	Nipissing	Geol. Surv.
Kapikwabikak lake	1,215	Rainy River	C. P. R.
Kashabowi		Thunder Bay	
Station	1,529		C. N. R.
Kashabowi riv 0·7 mile, E., water, 1,500 ; rail	1,517		C. N. R.
Kashabowi lake	1,263	Rainy River	C. P. R.
Kashishibog lake	1,474	Thunder Bay	C. P. R.
Katrine		Parry Sound	
Station	986		G. T. R.
Magapatawan river, 0·9 mile, S., water, 969 ; rail	987		G. T. R.
Kawagamong lake, high water, 664 ; low water	657	Parry Sound	C. P. R.
Kawashkagama lake	1,069	Thunder Bay	Geol. Surv.
Kawipetikwiwa lake	920	Thunder Bay	Geol. Surv.
Kearney	1,110	Parry Sound	C. A. R.
Kebsquashesing lake, high water, 1,407 ; low water	1,403	Algoma	C. P. R.
Keene		Peterborough	
Station	635		G. T. R.
Indian river, water, 629 ; rail	636		G. T. R.
Keene Summit		Peterborough	
Siding	752		G. T. R.
Summit, 0·1 mile, E., ground, 762 ; rail	754		G. T. R.
Keewatin	1,078	Rainy River	C. P. R.
Kekekkwabek lake	1,242	Rainy River	C. P. R.
Kelly lake	869	Nipissing	C. C. Co.
Kemptville junction		Grenville	
Station	330		C. P. R.
South Rideau river, 2·0 miles, W., water, 302 ; rail	322		C. P. R.
Kemptville station	297	Grenville	C. P. R.
Kendry	651	Peterborough	C. P. R.
Kepelee	660	Frontenac	C. P. R.
Kenilworth	1,481	Wellington	C. P. R.
Kennabutch	644	Algoma	C. P. R.
Kennabutch lake	1,337	Rainy River	C. P. R.
Kennedy lake	1,242	Rainy River	C. P. R.
Kenogami or Long lake	796	Algoma	C. P. R.
Kent Bridge	607	Kent	C. P. R.
Kerwood	768	Middlesex	G. T. R.
Kestrel lake	1,298	Thunder Bay	C. P. R.
Kettle lake	1,015	Nipissing	Geol. Surv.
Killaloe		Renfrew	
Station	601		C. A. R.
Brennan brook, water, 593 ; rail	602		C. A. R.
Killarney peak	1,386	Algoma	Admir. chart
Killenbeck lake	273	Leeds	Rys. & Canals
Kilworthy	755	Muskoka	G. T. R.
Kimball	629	Lambton	M. C. R.
Kinburn	314	Carleton	C. A. R.
Kincardine	589	Bruce	G. T. R.
King		York	
Station	957		G. T. R.
Summit, 3·9 miles, N.	1,020		G. T. R.
Kingscourt	709	Lambton	G. T. R.
Kingsmill	796	Elgin	M. C. R.
King Point bluff	951	Bruce	Admir. chart

ONTARIO

Locality	Elev	County	Authority
Kingston		Frontenac	G. T. R.
Grand Trunk station (Kingston)	249		G. T. R.
Grand Trunk station (Kingston junction)	273		G. T. R.
K. & P. station	255		K. & P. R.
Kingston & Pembroke Ry. crossing	287		G. T. R.
Bench marks—			
City hall	261·3		City Eng.
Corner of Johnston and Ontario sts	263·5		City Eng.
St. George church	277·1		City Eng.
Convent; Bagot street	263·5		City Eng.
Murney tower	264·0		City Eng.
Court house	300·1		City Eng.
Corner of Bagot and West streets	280·7		City Eng.
Gate post (Losher's), Princess st	352·9		City Eng.
Church, corner of Sydenham and William streets	304·1		City Eng.
Church, corner of Clergy and Johnston streets	316·6		City Eng.
Water tower base, 354 ; top	434		City Eng.
City hall, floor level	266·4		City Eng.
Highest point, near intersection of Princess st. and concession road	353		City Eng.
City datum	247·3		City Eng.
B.M., window sill, Carruthers' old store	249·2		D. W. Com.
Graving dock, coping	251·4		D. W. Com.
Graving dock, zero of gauge	244·8		Pub. Works
Kingsville	608	Essex	L. E. & D. R. R.
Kinmount station	919	Victoria	G. T. R.
Kinmount junction	935	Haliburton	G. T. R.
Kinogama		Algoma	
Station	1,370		C. P. R.
Apiskinagama river, 4·3 miles, E., water, 1,353 ; rail	1,364		C. P. R.
Summit, 2·8 miles, E	1,412		C. P. R.
Kippen	884	Huron	G. T. R.
Kirkfield	900	Victoria	G. T. R.
Kishkutena ridge	1,600	Rainy River	Geol. Surv.
Kitchiwano lake, water (Oct. 30, 1895)	766	Peterborough	Rys. & Canals
Kitchiwatchi lake	1,438	Thunder Bay	C. P. R.
Kleinburg	709	York	C. P. R.
Klock	530	Nipissing	C. P. R.
Knife lake	1,380	Rainy River	Dawson
Knowlton lake	449	Frontenac	K. & P. R.
Komoka		Middlesex	
Can. Pac. station	812		C. P. R.
Grand Trunk station	808		G. T. R.
Grand Trunk Ry., crossing	795		C. P. R.
Kukaganing lake	882	Nipissing	Geol. Surv.
Kukukus lake	1,281	Rainy River	C. P. R.
Kuminitikuchink lake	1,361	Rainy River	Dawson
Lac Poulin		Algoma	
Station	1,504		C. P. R.
Summit, 0·8 mile, E., ground, 1,517 ; rail	1,513		C. P. R.
Lac Poulin, water	1,481	Algoma	C. P. R.
Lady Evelyn lake	930	Nipissing	Geol. Surv.
Lady Macdonald lake	932	Nipissing	C. C. Co.
Lafontaine hill	1,040	Simcoe	Admir. chart
Lafontaine lake, high water, 299 ; low water	291	Renfrew	Rys. & Canals
Lake junction	580	Simcoe	G. T. R.
Lakefield station	770	Peterborough	G. T. R.
Lakefield junction	654	Peterborough	G. T. R.
L'Amable		Nipissing	
Station	1,127		C. A. R.
Madawaska river, 1·6 mile, W., water, 1,177 ; rail	1,188		C. A. R.
Lambton		York	
Station	399		C. P. R.
Humber river, high water, 312 ; low water, 303 ; rail	399		C. P. R.
Lancaster		Glengarry	
Station	164		G. T. R.
Door-sill of post office	165·0		D. W. Com.
B. M., north-west corner of station	165·0		D. W. Com.
B. M., west abutment of Black River bridge	165·3		D. W. Com.

ONTARIO

Locality	Elev.	County	Authority
Lansdowne............		Leeds ..	
Station.	333		G. T. R.
B. M., coping of culvert, half mile, W., of station...	325·0		D. W. Com.
Larchwood...		Algoma	
Station	868		C. P. R.
Vermilion river, water, 842; rail	867		C. P. R.
Larkin	551	Hastings	B. of Q. R.
La Seine.	1,174	Rainy River	C. N. R.
Laurel....	1,618	Dufferin	C. P. R.
La Vallee	1,126	Rainy River	C. N. R.
Lavant	845	Lanark	K. & P. R.
La Vase lake..........	772	Algoma	Geol. Surv.
Lawrence...........	724	Lambton	G. T. R.
Lawrence...	740	Elgin	G. T. R.
Leading-mark hill......	1,460	Algoma	Admir. chart
Leamington		Essex	
Mich. Cent. station........	626		M. C. R.
L. E. & D. R. station.........	624		L. E. & D. R. R.
Lake Erie and Detroit River Ry. crossing........	624		M. C. R.
Leaside.. ...		York	
Station........	429		C. P. R.
Don river, 1·1 mile, E., bed, 310 ; rail	417		C. P. R.
Lee......	399	Leeds	B. & W. R.
Lefroy......	770	Simcoe	G. T. R.
Leonard...	272	Russell	C. P. R.
Leslie.		Wentworth	
Station....	986		C. P. R.
Summit, 1·5 mile, E.....	1,003		C. P. R.
Lewisville (now Northwood).............	613	Kent	G. T. R.
Lily lake.....	1,001	Nipissing	Geol. Survey
Lily lake.....	1,285	Rainy River	C. P. R.
Limehouse..	1,002	Halton	G. T. R.
Lindsay.......		Victoria	
Station.......	854		G. T. R.
Scugog river bridge, rail......	859		G. T. R.
Scugog river, below lock, 813 ; above lock	821		Rys. & Canals
Linkooping		Thunder Bay	
Station	1,534		C. P. R.
South-east branch Savanne river, 3·3 miles, S., water, 1,538 ; rail	1,546		C. P. R.
Lion's end....	814	Bruce	Admir. chart
Lion Head....	1,083	Algoma	Admir. chart
Lion Rump........	1,050	Algoma	Admir. chart
Lisgar ...	676	Halton	C. P. R.
Lisle	743	Simcoe	G. T. R.
Listowel..	1,262	Perth	G. T. R.
Little lake......	620	Peterborough	Rys. & Canals
Little Buckhorn lake..	795	Peterborough	Rys. & Canals
Little Joe lake........	604	Lanark	K. & P. R.
Little Mud lake....		Algoma	
Standard high water	585·1		D. W. Com.
Standard low water..	580·3		D. W. Com.
Mean water, 1871 to 1900......	581·6		D. W. Com.
Little Sandy lake..	1,286	Rainy River	C. P. R.
Little Wabigoon or Dinorwic lake..	1,206	Rainy River	C. P. R.
Lobo Townline	860	Middlesex	G. T. R.
Lochalsh..	1,182	Algoma	C. P. R.
Lochlin....	883	Haliburton	G. T. R.
Locksley		Renfrew	
Station ...	513		C. A. R.
Ledgerwood brook, 0·7 mile, S., water, 477 ; rail...	503		C. A. R.
Locust Hill station......	667	York	C. P. R.
Lofoden (closed)....	1,072	Thunder Bay	C. P. R.
Londesborough.......		Huron	
Station........	971		G. T. R.
Maitland river, bed, 957 ; rail......	967		G. T. R.
London ...		Middlesex	
Grand Trunk station...	805		G. T. R.
Can. Pac. station.	804		C. P. R.
Asylum station..........	854		C. P. R.

ONTARIO

Locality	Elev.	County	Authority
London....................................		Middlesex	
London Yard............................	816		C. P. R.
Burwell street..........................	814		G. T. R.
Thames river, at junction of N. and S. branches, extreme high water (July 1881) 762; flood level, 867; normal low normal..................	753		City Eng.
Cove of Thames river, 1·5 mile, W., bed, 748; rail..	783		G. T. R.
North branch Thames river, 0·5 mile, W., bed, 753; rail....................................	796		G. T. R.
Thames river, 0·4 mile, S., bed, 771; rail..........	803		L. E. & D. R. R.
North branch Thames river,0·5 mile, W.,bed,760; rail.			
Springbank reservoir (16 feet deep) bottom........	987		City Eng.
City hall, floor level.....................	899		City Eng.
Highest point, corner St. James and Richmond sts.	940		City Eng.
London East....................................		Middlesex	
Southern division station.............	819		G. T. R.
St. Marys branch station..............	830		G. T. R.
Long Branch................................	310	York	G. T. R.
Long or Kenogami lake.................	796	Algoma	C. P. R.
Long lake...................................	1,281	Nipissing	C. A. R.
Long lake...................................	1,013	Thunder Bay	C. P. R.
Long lake, Little..........................	1,082	Thunder Bay	C. P. R.
Long lake...................................	271	Leeds.	Rys. & Canals
Long lake (Olden).........................	614	Frontenac	K. & P. R.
Long lake (Portland)......................	446	Frontenac	K. & P. R.
Long lake...................................	598	Lennox & Add'ton	Geol. Surv.
Long lake...................................	826	Parry Sound	C. P. R.
Long lake...................................	735	Ontario	G. T. R.
Longford...................................	1,098	Parry Sound	Geol. Surv.
Longwood...................................		Middlesex	
Can. Pac. station.....................	735		C. P. R.
Grand Trunk station...................	747		G. T. R.
Lyon..		Thunder Bay	
Station..................................	1,028		C. P. R.
Mackenzie river, 1·0 mile, W., water 989; rail.....	1,000		C. P. R.
Summit, 1·5 mile around, 1,099; rail...........	1,072		C. P. R.
Loon lake...................................	1,481	Thunder Bay	C. P. R.
Loop Line (now Sin........................	741	Norfolk	G. T. R.
Lorne Park.................................	308	Peel	G. T. R.
Lorneville...................................		Victoria	
Station..................................	892		G. T. R.
Coboconk branch.......................	860		G. T. R.
Loughborough lake........................	401	Frontenac	K. & P. R.
Lovesick lake..............................	790	Peterborough	Rys. & Canals
Lowbanks...................................	578	Haldimand	G. T. R.
Lucan......................................	980	Middlesex	G. T. R.
Lucan Crossing, London & Wingham branch, rail, 860; main line, rail..............................	881	Middlesex	G. T. R.
Lucknow....................................	908	Bruce	G. T. R.
Lune, Lac de...............................	964	Thunder Bay	C. P. R.
Lyn...		Leeds	
Grand Trunk station....................	284		G. T. R.
B. & W. station........................	291		B. & W. R.
B. M., coping of west abutment of small bridge, (G. T. Ry.) near station....................	285·5		D. W. Com.
Lynden......................................	751	Wentworth	G. T. R.
Lyndhurst....................................		Leeds	
Station..................................	297		B. & W. R.
Upper mill-pond........................	296		Rys. & Canals
Lower mill-pond........................	271		Rys. & Canals
Lynn Valley.................................	657	Norfolk	G. T. R.
Lynx lake...................................	1,025	Nipissing	K. & P. R.
Mabella.....................................	1,444	Thunder Bay	C. N. R.
Maberley....................................	576	Lanark	C. P. R.
Macaulay Central junction.................	1,053	Nipissing	C. A. R.
Macdonald or Cooper lake.................	914	Nipissing	Geol. Surv.
Mackenzie lake.............................	956	Nipissing	Geol. Surv.
Mackenzie....................................		Thunder Bay	
Station.................................	911		C. P. R.
Mackenzie river, water, 893; rail...........	902		C. P. R.

ONTARIO

Locality	Elev.	County	Authority
Mackey..	431	Renfrew	C. P. R.
MacMillan	1,238	Rainy River	C. P. R.
Madawaska		Nipissing	
Station....	1,035		C. A. R.
Opeongo river, 6.1 miles, E., water, 1,069; rail....	1,080		C. A. R.
Madawaska river, water, 1,014; rail...........	1,031		C. A. R.
Madoc junction	502	Hastings	G. T. R.
Madoc station....	575	Hastings	G. T. R.
Madrid P.O, ...	1,449	Renfrew	K. & P. R.
Magpie lake.	1,138	Algoma	C. P. R.
Maidstone Cross,.	627	Essex	M. C. R.
Maitland		Grenville	
Station....	327		G. T. R.
B. M., west abutment of culvert, half mile W.. ...	340.9		D. W. Com.
Makadawaganus lake..	1,265	Algoma	C. P. R.
Makoping or Windy lake................	1,080	Algoma	Geol. Surv.
Malcolm bluff.	963	Bruce	Admir. chart
Mallorytown		Leeds	
Station.	335		G. T. R.
B. M., end of box culvert, half mile W...........	325.7		D. W. Com.
Malone..		Hastings	
Station....	718		C. O. R.
Moira river, bed, 708; rail...........	718		C. O. R.
Summit, 2.3 miles, W..............	727		C. O. R.
Malton	549	Peel	G. T. R.
Mamainse hill	1,850	Algoma	Admir. chart
Manchester..............	955	Ontario	G. T. R.
Mandaunin	645	Lambton	G. T. R.
Manilla.....................	964	Victoria	G. T. R.
Manitou lake.................	1,215	Rainy River	C. P. R.
Manitonnamaig lake....	1,033	Thunder Bay	Geol. Surv.
Manitowik lake...	996	Algoma	C. P. R.
Mann lake.............	975	Nipissing	Geol. Surv.
Mannajigame lake............	1,075	Nipissing	Geol. Surv.
Manotick..,..............	327	Carleton	C. P. R.
Mansewood.............	732	Halton	G. T. R.
Manvers.............		Durham	
Station..............	958		C. P. R.
Grand Trunk Ry. crossing ,3.5 miles E., G. T.Ry., rail, 840; C. P. Ry., rail..............	865		C. P. R.
Maple...	817	York	G. T. R.
Maple bluff..	842	Algoma	Admir. chart
Maple island, water in Maganatawan river...	797	Parry Sound	Geol. Surv.
Maple Lake station	798	Parry Sound	C. A. R.
Maple lake.............	746	Parry Sound	C. A. R.
Marden	1,140	Wellington	G. T. R.
Margach (formerly Rossland)...........	1,129	Rainy River	C. P. R.
Mariposa....	866	Victoria	G. T. R.
Markdale..	1,357	Grey	C. P. R.
Markham Road	533	York	G. T. R.
Markham Road	564	York	C. P. R.
Markhamville...........	640	York	G. T. R.
Markstay....	685	Nipissing	C. P. R.
Marlbank.................	537	Hastings	B. of Q. R.
Marmora		Hastings	
Cent. Ont. station..........	594		C. O. R.
Ont. Bel. & Nor. station...........	599		O. B. & N. R.
Marsh lake..............	746	Parry Sound	C. A. R.
Marshall Bay station...	312	Carleton	C. A. R.
Marshfield.......	608	Essex	L. E. & D. R. R.
Marsh Hill station.......	863	Ontario	G. T. R.
Marshville.............	580	Welland	G. T. R.
Marten lake..............	936	Nipissing	Geol. Surv.
Martin.		Rainy River	
Station	1,558		C. P. R.
Summit, 0.4 mile, E., ground, 1,569; rail........	1,559		C. P. R.
Mary lake.	928	Muskoka	Geol. Surv.
Marysville..............	335	Hastings	G. T. R.
Masinanwaning, Lake..............	900	Algoma	C. P. R.
Maskinonge-wagami lake............	836	Nipissing	Geol. Surv.

ONTARIO

Locality	Elev.	County	Authority
Massey	667	Algoma	C. P. R.
Matawanan lake	1,430	Algoma	C. P. R.
Matawin	1,291	Thunder Bay	C. N. R.
Mattagamashing lake	871	Nipissing	Geol. Surv.
Mattawa		Nipissing	
Station	564		C. P. R.
Town Hall station	521		C. P. R.
Ottawa river, high water (1890), 510 ; low water, 496 ; rail	521		C. P. R.
Maxville		Glengarry	
Station	335		C. A. R.
Summit, 1·9 mile, E.	362		C. A. R.
Maxwell	1,023	Haliburton	
Station	995		I. B. & O. R.
Burnt river, 6·5 mile, W., bed, 974 ; rail			I. B. & O. R.
Mazininwaning or Vermilion lake	786	Algoma	Geol. Surv.
Mazokama	648	Thunder Bay	C. P. R.
McAlpin	219	Prescott	C. P. R.
McBean mountain	1,207	Algoma	Admir. chart
McDiarmid lake	992	Nipissing	Geol. Surv.
McGregor		Essex	
Can. Pac. station	955		C. P. R.
Mich. Cent. station	608		M. C. R.
Lake Erie & Detroit River Ry. crossing	609		M. C. R.
McKay lake	1,047	Thunder Bay	Geol. Surv.
McKay mountain	1,600	Thunder Bay	Admir. chart
McLean	749	Thunder Bay	C. P. R.
McQuaig hill (Cockburn island)	1,060	Algoma	Admir. chart
McRae	976	Wentworth	C. P. R.
Meadowside	661	Nipissing	C. P. R.
Meadowvale	563	Peel	C. P. R.
Meaford	672	Grey	G. T. R.
Megrund	1,429	Rainy River	C. P. R.
Melancthon	1,660	Dufferin	C. P. R.
Melbourne	735	Middlesex	M. C. R.
Melgund		Thunder Bay	
Station	817		C. P. R.
Black Pic river, 0·8 mile, W., we 6 ; rail	805		C. P. R.
Melrose	867	Middlesex	C. P. R.
Melville junction	1,326	Peel	C. P. R.
Merlin	634	Kent	L. E. & D. R. R.
Merrickville		Lanark	
Station	352		C. P. R.
Rideau canal, water, 308 ; rail	352		C. P. R.
Merritton		Lincoln	
Station	389		G. T. R.
Welland canal tunnel, 2·0 miles, E., rail	448		G. T. R.
Metagama		Algoma	
Station	1,274		C. P. R.
West branch Spanish river, 4·0 miles, W., water, 1,267 ; rail	1,273		C. P. R.
Metcalfe village	286	Carleton	T. & O. R.
Meyersburg village	403	Northumberland	C. N. & P. R.
Michipicoten		Algoma	
Ore dock	641		A. C. & H. B. R.
Commercial dock	607		A. C. & H. B. R.
Michipicoten island	537	Thunder Bay	Admir. chart
Middlemiss	709	Middlesex	G. T. R.
Middleport	662	Brant	G. T. R.
Middleton Line	841	Oxford	G. T. R.
Middleton		Thunder Bay	
Station	691		C. P. R.
Little Pic river, 1·8 mile, E., water, 612 ; rail	693		C. P. R.
Midland	583	Simcoe	G. T. R.
Mildmay	1,028	Bruce	G. T. R.
Mile lake	612	Renfrew	K. & P. R.
Milieu or Savanne lake	1,536	Thunder Bay	Dawson
Millbridge		Hastings	
Station	944		C. O. R.
Summit, 0·6 mile, N.	960		C. O. R.

ONTARIO

Locality	Elev.	County	Authority
Millbrook		Durham	
Station	768		G. T. R.
Summit, 2·1 miles, N.	879		G. T. R.
Mille Lacs, Lac des	1,496	Thunder Bay	Dawson
Miller lake	977	Nipissing	Geol. Surv.
Mille Roches		Stormont	
Station	225		G. T. R.
B. M., west wall of culvert, half mile W. of station	227·5		D. W. Com.
Milliken		York	
Station	659		G. T. R.
Summit, 0·7 mile, N.	665		G. T. R.
Milton		Halton	
Can. Pac. station	660		C. P. R.
Grand Trunk station	661		G. T. R.
Grand Trunk Ry. crossing	661		C. P. R.
Mimico	300	York	G. T. R.
Mine Centre	1,194	Rainy River	C. N. R.
Mineral Springs	581	Wentworth	T. H. & B. R.
Minesing		Simcoe	
Station	619		G. T. R.
Summit, 1·4 mile, N.	758		G. T. R.
Mingush lake	1,454	Thunder Bay	C. P. R.
Mink lake	1,427	Rainy River	C. N. R.
Minnebaha lake	1,296	Rainy River	C. P. R.
Minnitaki	1,139	Rainy River	C. P. R.
Minnitaki lake	1,154	Rainy River	C. P. R.
Missinaibi		Algoma	
Station	1,106		C. P. R.
Summit, 2·2 miles, W., ground, 1,106; rail	1,140		C. P. R.
Missinaibi lake	1,134	Algoma	C. P. R.
Mississippi		Frontenac	
Station	690		K. & P. R.
Mississippi river, 1·3 mile, N., water, 647; rail	661		K. & P. R.
Mitchell		Perth	
Station	1,121		G. T. R.
North branch, Thames river, water, 1,085; rail	1,122		G. T. R.
Moffat	1,047	Wellington	C. P. R.
Moira lake	510	Hastings	G. T. R.
Mokoman	1,003	Thunder Bay	C. N. R.
Monklands	328	Stormont	C. P. R.
Mono Road	973	Peel	C. P. R.
Montague	583	Welland	M. C. R.
Montizambert		Thunder Bay	
Station	1,090		C. P. R.
White river, 1·9 mile, W., high water, 1,073; low water, 1,068; rail	1,081		
Montrose station	595	Welland	M. C. R.
Montrose junction	612	Welland	M. C. R.
Monument lake	1,178	Rainy River	C. P. R.
Moon lake	1,097	Algoma	A. C. & H. B. R.
Moon lake	1,313	Nipissing	C. P. R.
Moorefield	1,349	Wellington	G. T. R.
Moore Point	606	Algoma	A. C. & H. B. R.
Mooretown		Lambton	
Station	606		L. E. & D. R. R.
Summit, 1·5 mile, N.	638		L. E. & D. R. R.
Moore lake	613	Renfrew	C. P. R.
Moorlake		Renfrew	
Station	623		C. P. R.
Summit, 2·8 miles, S., ground, 710; rail	696		C. P. R.
Moose Creek station	288	Stormont	C. A. R.
Moose lake (Seine river)	1,342	Rainy River	Dawson
Moose lake (Pigeon river)	1,490	Thunder Bay	Dawson
Moosewater lake	1,288	Algoma	C. P. R.
Morrisburg		Dundas	
Station	268		G. T. R.
B. M., west wall of culvert, half mile E. of station	265		D. W. Com.
B. M., on station	273		D. W. Com.

ONTARIO

Locality	Elev.	County	Authority
Mosborough	1,084	Wellington	G. T. R.
Moscow	468	Lennox & Add'ton	B. of Q. R.
Mosquito lake	403	Leeds	Rys. and Canals
Moule lake	1,442	Algoma	C. P. R.
Moulton	587	Haldimand	G. T. R.
Mountain	273	Dundas	C. P. R.
Mountain or Round lake	911	Nipissing	Geol. Surv.
Mountain lake	1,029	Nipissing	Geol. Surv.
Mountain or Bass lake	1,332	Renfrew	K. & P. R.
Mountain lake (Pigeon river)	1,651	Thunder Bay	Dawson
Mountain lake (Kenogami river)	1,100	Thunder Bay	Geol. Surv.
Mountain Grove		Frontenac	
Station	681		C. P. R.
Summit, 2·2 miles, E., ground, 769; rail	749		C. P. R.
Mount Albert station	795	York	G. T. R.
Mount Brydges station	821	Middlesex	G. T. R.
Mount Forest station	1,348	Wellington	C. P. R.
Mount Forest junction	1,378	Wellington	C. P. R.
Mount Pleasant		Brant	
Grand Trunk station	800		G. T. R.
T. H. & B. station	790		T. H. & B. R.
Mount Vernon station	801	Brant	G. T. R.
Moxam lake	933	Nipissing	Geol. Surv.
Mud lake	1,325	Rainy River	C. P. R.
Mud lake	403	Leeds	Rys. & Canals
Mud lake	900	Nipissing	Geol. Surv.
Mud lake (Bedford)	448	Frontenac	K. & P. R.
Mud lake (Portland)	446	Frontenac	K. & P. R.
Mud lake	448	Lennox & Add'ton	B. of Q. R.
Mud lake	1,488	Thunder Bay	C. N. R.
Mud lake (Pic-river)	1,050	Thunder Bay	Geol. Surv.
Mud or Watap lake (Pigeon river)	1,667	Thunder Bay	Dawson
Mud lake (Westmeath)	409	Renfrew	C. P. R.
Mud lake (Bonnechere river)	550	Renfrew	Geol. Surv.
Mudlake Bridge	453	Lennox & Add'ton	K. &
Mud Lake siding	1,138	Parry Sound	C. A.
Muddy lake	1,045	Thunder Bay	Geol. S
Muirkirk		Kent	
Mich. Cent. station	722		M. C. R.
L. E. & D. R. station	728		L. E. & D. R. R.
Mull	647	Kent	M. C. R.
Mumford		Haliburton	
Station	1,259		I. B. & O. R.
Summit, 1·1 mile, E	1,289		I. B. & O. R.
Muncey		Middlesex	
Station	726		M. C. R.
Thames river, 1·3 mile, E., bed, 645; rail	720		M. C. R.
Munro	1,098	Hastings	C. O. R.
Murdoch		Thunder Bay	
Station	877		C. N. R.
Whitefish river, 2·4 miles, W., bed, 935; rail	943		C. N. R.
Murillo		Thunder Bay	
Station	947		C. P. R.
Summit, 3·0 miles, N	1,080		C. P. R.
Murphy		Grey	
Station	796		G. T. R.
Summit, 1·6 mile, W	826		G. T. R.
Murray Hill		Northumberland	
Station (closed)	346		G. T. R.
Summit, 0·3 mile, E	354		G. T. R.
Murray lake	815	Nipissing	Geol. Surv.
Murvale	470	Frontenac	K. & P. R.
Muskoka lake	745	Muskoka	Interior
Muskoka Wharf	749	Muskoka	G. T. R.
Muskrat lake	409	Renfrew	C. P. R.
Myrtle		Ontario	
Can. Pac. station	888		C. P. R.

ONTARIO

Locality	Elev.	County	Authority
Myrtle.....		Ontario	
Grand Trunk station.....	821		G. T. R.
Grand Trunk crossing, G. T. Ry., rail, 895 ; Can. Pac. Ry., rail	870		C. P. R.
Nairn		Algoma	
Station.....	719		C. P. R.
Spanish river, 3·7 miles, W., water, 634 ; rail.....	670		C. P. R.
Naked lake.....	764	Parry Sound	C. P. R.
Nanakan lake.....	1,117	Rainy River	Dawson
Napanee lake.....	446	Frontenac	K. & P. R.
Napanee		Lennox & Add'ton	
Station	313		G. T. R.
Napanee river, water (June 26, 1878), 276 ; rail.....	312		G. T. R.
Napanee Mills.....		Lennox & Add'ton	
Station	310		B. of Q. R.
Napanee river, 2·1 miles, W., water, 274 ; rail.....	285		B. of Q. R.
Narrow lake.....	1,235	Rainy River	C. P. R.
Nasbonsing.....	785	Nipissing	C. P. R.
Nassau Mills.....	708	Peterborough	G. T. R.
Natamasagami lake..	1,107	Thunder Bay	C. P. R.
Naughton	804	Algoma	C. P. R.
Navan.....	240	Russell	C. P. R.
Neebing.....	715	Thunder Bay	C. P. R.
Nelles Corners.....	713	Haldimand	G. T. R.
Nemegosenda.....		Algoma	
Station	1,426		C. P. R.
Nemegosenda river, water, 1,419 ; rail.....	1,425		C. P. R.
Summit, 2·9 miles, E., ground, 1494 ; rail.....	1,487		C. P. R.
Nequaquon, Lake or Lac la Croix.....	1,181	Rainy River	Dawson
Net lake.....	965	Nipissing	Geol. Surv.
Newboro.....	437	Leeds	B. & W. R.
Newburg.....	334	Lennox & Add'ton	B. of Q. R.
Newbury.....	701	Middlesex	G. T. R.
New Canaan.....	608	Essex	L. E. & D. R. R.
Newcastle.....	295	Durham	G. T. R.
New Durham.....	836	Brant	G. T. R.
Newington.....	321	Stormont	N. Y. & O. R.
New Lowell.....	637	Simcoe	G. T. R.
Newmarket.....	772	York	G. T. R.
New Sarum.....	759	Elgin	G. T. R.
Newtonville.....	392	Durham	G. T. R.
New Toronto.....	316	York	G. T. R.
Niagara Falls.....		Welland	
Grand Trunk station.....	573		G. T. R.
Mich. Cent. station.....	609		M. C. R.
Niagara river, water, 342 ; rail.....	573		G. T. R.
Niagara junction.....	606	Welland	M. C. R.
Niagara-on-the-lake.....		Lincoln	
Station	250		M. C. R.
Queen st.....	287		M. C. R.
Chatauqua switch.....	289		M. C. R.
Paradise switch.....	308		M. C. R.
Niblock.....		Thunder Bay	
Station.....	1,536		C. P. R.
Fire-steel river, 2·7 miles, E., water, 1,505 ; rail..	1,513		C. P. R.
Niles Corners.....	312	Prince Edward	C. O. R.
Nipigon.....		Thunder Bay	
Station.....	685		C. P. R.
Summit, 1·9 mile, E., ground, 728 ; rail.....	713		C. P. R.
Depression, 2·7 miles, W., rail.....	607		C. P. R.
Nipigon river bridge, rail.....	685		C. P. R.
Nipigon, Lake.....	852	Thunder Bay	C. P. R.
Nipigon hills.....	1,600	Thunder Bay	Admir. chart
Nipissing junction.....	670	Nipissing	C. P. R.
Nipissing, Lake, high water, 648 ; low water.....	641	Nipissing	Rys. & Canals
Nitawagami or Whitewater lake.....	835	Algoma	Geol. Surv.
Nixon.....	772	Norfolk	G. T. R.
Nonwakaming lake.....	955	Nipissing	Geol. Surv.
Nonwatan lake (Black Sturgeon river).....	738	Thunder Bay	C. P. R.
Nonwatin lake.....	1,238	Rainy River	Dawson

ONTARIO

Locality	Elev.	County	Authority.
Nordland (closed)	1,543	Thunder Bay	C. P. R.
Norman	1,106	Rainy River	C. P. R.
Norris lake	935	Nipissing	Geol. Surv.
North Bay	659	Nipissing	C. P. R.
North Bothwell	661	Lambton	C. P. R.
Northeast hill	1,760	Algoma	Admir. chart
Northern Light lake	1,441	Thunder Bay	Crown Lands
Northfield		Stormont	
Station	331		N. Y. & O. R.
Summit, 0·3 mile, E., ground, 340; rail	334		N. Y. & O. R.
North Glencoe		Middlesex	
Station	731		C. P. R.
Grand Trunk Ry. crossing, 1·7 mile, W	718		C. P. R.
North lake	1,549	Thunder Bay	C. N. R.
North Lake		Thunder Bay	
Station	1,572		C. N. R.
Summit, 6·7 miles E.	1,692		C. N. R.
North Newbury	681	Middlesex	C. P. R.
North Parkdale	303	York	G. T. R.
North Thamesville	622	Kent	C. P. R.
North Toronto	400	York	C. P. R.
Northwood	613	Kent	G. T. R.
Norval	818	Halton	G. T. R.
Norway lake	545	Renfrew	K. & P. R.
Norwich		Oxford	
B. & T. station	846		G. T. R.
G. B. & L. E. station	873		G. T. R.
Norwich junction	842	Oxford	G. T. R.
Norwood	671	Peterborough	C. P. R.
Nosbonsing Crossing	745	Parry Sound	G. T. R.
Nosbonsing lake, high water, 781; mean water	776	Nipissing	Geol. Surv.
Nottawa	707	Simcoe	G. T. R.
Novar	1,075	Parry Sound	G. T. R.
Oakland	635	Essex	M. C. R.
Oakville	329	Halton	G. T. R.
Oba lake	1,218	Algoma	C. P. R.
Obabica lake	932	Nipissing	Geol. Surv.
O'Brien		Algoma	
Station (summit)	1,442		C. P. R.
White river, 2·3 miles E., water, 1.302; rail	1,331		C. P. R.
Ocean House	256	Wentworth	G. T. R.
Oil City	661	Lambton	M. C. R.
Oiseau Bay hill	1,500	Thunder Bay	Admir. chart
Oldcastle	627	Essex	L. E. & D. R. R.
Olden	676	Frontenac	K. & P. R.
Old Ferry	727	Thunder Bay	C. N. R.
Old Fort	589	Simcoe	G. T. R.
Olier lake	984	Nipissing	Geol. Surv.
Omemee		Victoria	
Station	857		G. T. R.
Pigeon creek, water (Aug. 12, 1899), 807; rail	825		G. T. R.
Ompah		Frontenac	
Station	814		K. & P. R.
Antoine brook, water, 776; rail	787		K. & P. R.
Onaping station (closed)	1,047	Algoma	C. P. R.
Onaping lake	1,417	Algoma	Geol. Surv.
Onaping lake, Lower	1,410	Algoma	Geol. Surv.
Onondaga	663	Brant	G. T. R.
Ontario, Belmont & Northern junction	598	Hastings	O. B. & N. R.
Ontario, Lake			
Standard high water, (extreme high water, May,1870)	248·7		D. W. Com.
Extreme low water, (March, 1825)	242·1		D. W. Com.
Standard low water	244·2		D. W. Com.
Mean water, (1860 to 1900)	246·0		D. W. Com.
Mean water, (1871 to 1900)	245·7		D. W. Com.
Opeongo	507	Renfrew	K. & P. R.
Opeongo and Aird roads, junction	725	Renfrew	K. & P. R.
Opeongo and Hyland lake roads, junction	777	Renfrew	K. & P. R.
Opeongo and Rockingham roads, junction	1,216	Renfrew	K. & P. R.

ONTARIO

Locality	Elev.	County	Authority
Opeongo and Trout lake roads, junction...........	686	Renfrew	K. & P. R.
Opinicon lake	396	Leeds	Rys. & Canals
Ops.............................	883	Victoria	G. T. R.
Orangeville station............................	1,395	Dufferin	C. P. R.
Orangeville junction..	1,613	Dufferin	C. P. R.
Ourang-outang lake........	1,339	Rainy River	C. P. R.
Orillia............................		Simcoe	
Station........	732		G. T. R.
Midland Ry. station (closed).............	724		G. T. R.
Ormsby station....	1,146	Hastings	C. O. R.
Ormsby junction....	1,160	Hastings	C. O. R.
Oro............................	790	Simcoe	G. T. R.
Orton....		Wellington	
Station	1,448		C. P. R.
Summit, 2·2 miles, W.......	1,528		C. P. R.
Osaquan		Rainy River	
Station	1,429		C. P. R.
Osaquan river, 1·8 mile, W., bed, 1,413 ; rail.... ...	1,423		C. P. R.
Osgoode...........................	301	Carleton	C. P. R.
Oshawa.......	331	Ontario	G. T. R.
Oskondiga.............		Thunder Bay	
Station....	1,419		C. P. R.
Oskondiga river, 2·1 miles, W., water, 1,428 ; rail..	1,441		C. P. R.
Osler bluff....	1,655	Grey	Admir. chart
Oso..		Frontenac	
Station.	674		K. & P. R.
South branch Bolton brook, 0·6 mile, N., water, 643 ; rail.... ....,	653		K. & P. R.
Ostersund	1,116	Rainy River	C. P. R.
Otonabee................................	674	Peterborough	C. P. R.
Ottawa.............................		Carleton	
Central station	212		C. A. R.
Union station......	178		C. P. R.
Sussex street station.......	190		C. P. R.
Elgin street station (closed)....	222		C. A. R.
Montreal & Ottawa junction......	198		C. P. R.
Chaudière junction...........	206		C. A. R.
Chaudière junction (Prescott branch).............	274		C. P. R.
Janeville, Montreal road crossing	187		C. P. R.
Parliament building, basement floor	284		Public Works
Parliament building, top of central tower........	440		Public Works
Eastern departmental building, basement floor.....	260		Public Works
Western departmental building, basement floor....	267		Public Works
Bench Marks—			
City Hall, floor level....	247		City Eng.
Waterworks, machinery bed of No. 1 set of pumps..........................	160		City Eng.
Old Men's Home, Bank street, water table.....	227		City Eng.
Cummings bridge, N. E. corner of wing wall...	192		City Eng.
Dyke, Dufferin road, W. end of..	183		City Eng.
Majors Hill park, stone fence post, corner Mackenzie ave. and Rideau street....	229		City Eng.
Athletic Club, Elgin street, basement window sill...........................	222		City Eng.
Sandy hill, corner Besserer and Wurtemburg sts...	233		City Eng.
Ashburnham hill, corner Maria and Concession sts.	261		City Eng.
Corner of Carling and Turner streets.	250		City Eng.
City datum, mitre sill of lower lock	129		C. P. R.
Alexandra bridge, rail....................	190		P. P. J. R.
Rideau canal........................			
Water above Ottawa locks	208		C. P. R.
Water above Hartwell locks.......	228		Rys. & Canals
Water above Hogsback lock...	242·5		Rys. & Canals
Rideau river........................			
At bridge near Sussex street station, high water, 177 ; rail....	190		C. P. R.
At Cummings bridge, flood (1896), 184 ; summer water........................	174		City Eng.
At Hurdman bridge, flood (1898), 192 ; low water.	180		Rys. & Canals

ONTARIO

Locality	Elev.	County	Authority
Ottawa......		Carleton	
Ottawa river..........................			
At foot of locks, extreme high water (May 16, 1896), 149·5 ; extreme low water (Sept. 28, 1881), 124·6 ; normal low water..........	128		Rys. & Canals
At Prince of Wales bridge, high water (1876)..	174		C. P. R.
Above Little Chaudière rapid., high water, 183 ; low water........	176		Rys. & Canals
Above Remoux rapids, high water, 187 : low water	179		Rys. & Canals
Otter..	1,210	Algoma	C. P. R.
Otter lake............................... . .	1,093	Algoma	A. C. & H. B. R.
Otter lake..	681	Parry Sound	C. A. R.
Otter lake	1,025	Renfrew	C. A. R.
Otter lake.	1,415	Thunder Bay	C. P. R.
Otter lake..	453	Frontenac	K. & P. R.
Otter Lake station......	732	Parry Sound	C. A. R.
Otter-slide lake..	1,340	Nipissing	Geol. Surv.
Ottertail lake........................	636	Algoma	B. M. & A. R.
Otterville........................	798	Oxford	G. T. R.
Ouimet	739	Thunder Bay	C. P. R.
Overholt........................	584	Welland	G. T. R.
Owen Sound......		Grey	
Grand Trunk station.....	585		G. T. R.
Can. Pac. station......	586		C. P. R.
Oxdrift........................	1,161	Rainy River	C. P. R.
Oxford....		Grenville	
Station.	352		C. P. R.
Summit, 3·8 miles, W., rail....	353		C. P. R.
Oxford Mills..............................	335	Grenville	C. P. R.
Ox-tongue lake.........	1,180	Haliburton	Geol. Surv.
Pagutchi lake..	1,371	Rainy River	C. P. R.
Paisley		Bruce	
Station	773		G. T. R.
Teeswater river, water, 709 ; rail..	770		G. T. R.
Pakeeshkon lake..	1,527	Thunder Bay	C. P. R.
Pakenham.......		Lanark	
Station	320		C. P. R.
Mississippi river, high water, 294 ; low water, 289 ; rail........	321		C. P. R.
Palgrave. .		Peel	
Station	931		G. T. R.
Summit, 1·0 mile, N., ground, 975 ; rail.........	969		G. T. R.
Palmerston	1,313	Perth	G. T. R.
Panache lake..	761	Algoma	Geol. Surv.
Papineau lake................	1,214	Haliburton	Geol. Surv.
Paquette....	616	Essex	L. E. & D. R.R.
Pardee		Algoma	
Station.....	1,528		C. P. R.
Summit, height-of-land between St. Lawrence and Hudson bay........	1,536		C. P. R.
Parham.....		Frontenac	
Station ...., ...	651		K. & P. R.
Eagle brook, 0·6 mile, N., water, 624 ; rail	649		K. & P. R.
Paris.		Brant	
Station...	843		G. T. R.
Grand river, 0·7 mile, S., water (Aug. 1878), 731; rail	813		G. T. R.
Grand river, 2·5 miles, E., bed, 743 ; rail..	803		G. T. R.
Parkdale....	305	York	C. P. R.
Parkhead....	723	Bruce	G. T. R.
Parkhill....	662	Middlesex	G. T. R.
Parry Sound (Rose Point)	602	Parry Sound	C. A. R.
Parrywood........	1,294	Rainy River	C. P. R.
Parrywood lake.....	1,375	Rainy River	C. P. R.
Partridge island..	770	Algoma	Admir. chart
Patton siding....	888	Oxford	G. T. R.
Paugon hill......	1,450	Thunder Bay	Admir. chart
Paul lake......	1,253	Nipissing	C. P. R.
Paulett, Cape..........	980	Bruce	Admir. chart

ONTARIO

Locality	Elev.	County	Authori
Payne........		Elgin	
Station....	747		G. T. R.
Dodd creek, bed, 720 ; rail	744		G. T. R.
Michigan Central crossing	742		G. T. R.
Pays Plat........	621	Thunder Bay	C. P. R.
Pearl..		Thunder Bay	
Station....	850		C. P. R.
Pearl river, water, 840; rail.....	851		C. P. R.
Peeshabo lake........	1,005	Nipissing	Geol. Surv.
Pelton....	627	Essex	M. C. R.
Pembroke...		Renfrew	
Can. Pac. station..	381		C. P. R.
Can. Atl. station... ..	393		C. A. R.
Bay of Ottawa river, high water (May, 1876), 375 ; low water (June 14, 1878), 368 ; rail........	381		C. P. R.
Muskrat river, 1·2 mile, S., water, 392 ; rail ..	418		C. A. R.
Pender lake.....	839	Parry Sound	C. A. R.
Pendleton..	234	Prescott	C. P. R.
Penetanguishene..		Simcoe	
Station	589		G. T. R.
Reformatory hill........	790		Admir. chart
Peninsula..	707	Thunder Bay	C. P. R.
Peninsula Harbour hill. .. '	1,400	Thunder Bay	Admir. chart
Peninsula lake....... ..	936	Muskoka	G. T. R.
Perch......	583	Lambton	G. T. R.
Perch lake.......	1,240	Rainy River	C. P. R.
Percy Reach	571	Northumberland	Rys. & Canals
Perkinsfield (summit)........	790	Simcoe	G. T. R.
Pero lake	553	Renfrew	K. & P. R.
Perry........	585	Welland	M. C. R.
Perrytown..	640	Durham	G. T. R.
Perth.....	433	Lanark	C. P. R.
Peterborough		Peterborough	
Grand Trunk station...	649		G. T. R.
Lakefield switch.........	656		G. T. R.
Can. Pac. crossing.....	638		G. T. R.
Can Pac. station....	533		C. P. R.
Little lake (Otonabee river), navigation level..	620		G. T. R.
Petersburg..		Waterloo	
Station....	1,210		G. T. R.
Summit, 1·0 mile W., ground, 1,260 ; rail........	1,248		G. T. R.
Peterson....	557	Hastings	C. P. R.
Petewawa..		Renfrew	
Station	467		C. P. R.
Petewawa river, water, 437 ; rail.	461		C. P. R.
Petrolea station	667	Lambton	G. T. R.
Petrolea junction.... . ..	716	Lambton	G. T. R.
Petrolea junction	690	Lambton	M. C. R.
Petrout lake....	996	Nipissing	Geol. Surv.
Pettit Mill.......	505	Essex	M. C. R.
Phelan	927	Algoma	C. P. R.
Phelpston..	724	Simcoe	G. T. R.
Pickerel lake.....	1,214	Rainy River	C. P. R.
Pickerel or Kaogassikok lake...	1,337	Rainy River	C. P. R.
Pickering.....	289	Ontario	G. T. R.
Picton.... .	322	Prince Edward	C. O. R.
Pie island... ..	1,450	Thunder Bay	Admir. chart
Pigeon lake...	805	Peterborough	Rys. & Canals
Pike Creek station.	588	Essex	G. T. R.
Pike lake, high water, 1,254 ; low water.......	1,249	Rainy River	C. N. R.
Pimisi or Eel lake, low water......	591	Nipissing	Rys. & Canals
Pine lake	869	Frontenac	K. & P. R.
Pine lake	891	Thunder Bay	Geol. Surv.
Pine lake (Kalmar)	1,184	Rainy River	C. P. R.
Pine lake (English river)	1,338	Rainy River	C. P. R.
Pinewood........	1,074	Rainy River	C. N. R.
Pinkerton.	859	Bruce	G. T. R.
Piperville.....	261	Carleton	N. Y. & O. R.

ONTARIO

Locality	Elev.	County	Authority
Plantagenet...............................		Prescott	
Station.....................	170		C. P. R.
South Nation river, high water (1862), 163·5; (March 17, 1898), 158·8; ordinary high water, 156·8; extreme low water, 147·9; rail......	170		C. P. R.
Plaunt hill........................	1,386	Renfrew	K. & P. R.
Plein Chant, Lac, low water.....	517	Nipissing	Rys. & Cana
Pogamasing....	1,164	Algoma	C. P. R.
Pogamasing lake..........................	1,201	Algoma	Geol. Surv.
Point Edward........................	585	Lambton	G. T. R.
Pontypool		Durham	
Station....	1,067		C. P. R.
Summit, 1·9 mile, W........	1,105		C. P. R.
Poplar lake........................	949	Nipissing	Geol. Surv.
Portage Road........	920	Victoria	G. T. R.
Port Arthur..........		Thunder Bay	
Can. Nor. station	607		C. N. R.
Can. Pac. station.....	608		C. P. R.
Port Britain....	256	Durham	G. T. R.
Port Burwell....	580	Elgin	T. L. E. & P. R.
Port Colborne.....		Welland	
Station	580		G. T. R.
Zero of gauge ...	580·4		D. W. Com.
Upper mitre sill...	557·3		D. W. Com.
B. M., on custom house ...	584·3		D. W. Com.
Port Coldwell (now Coldwell)........		Thunder Bay.	
Station....	711		C. P. R.
Summit, 3·1 miles, W., ground, 893; rail....	865		C. P. R.
Port Credit..............	206	Peel	G. T. R.
Port Dalhousie.....		Lincoln	
Station	249		G. T. R.
Zero of gauge	257·7		D. W. Com.
Lower mitre sill	239·9		D. W. Com.
Port Dover		Norfolk	
G. B. & L. E. station.	576		G. T. R.
H. & N. W. station	590		G. T. R.
Port Dover junction	784	Norfolk	G. T. R.
Port Elgin....	673	Bruce	G. T. R.
Port Hope station...	265	Durham	G. T. R.
Port Hope junction...	286	Durham	G. T. R
Port Lambton..	583	Lambton	L. E. & ..
Portlock....	502	Algoma	C. P. R.
Port Perry....	829	Ontario	G. T. R.
Port Robinson	586	Welland	G. T. R.
Port Rowan	622	Norfolk	G. T. R.
Port Stanley....	579	Elgin	L. E. & D. R.R.
Port Union....	265	Ontario	G. T. R.
Poshkokagan lake....	1,353	Thunder Bay	C. P. R.
Post hill....	1,005	Algoma	Admir. chart
Potabawinnana lake....	676	Parry Sound	C. A. R.
Potter lake....	1,428	Nipissing	C. A. R.
Pottersburg....	864	Middlesex	G T. R.
Poulin	1,504	Algoma	C. P. R.
Poulin, Lac....	1,481	Algoma	C. P. R.
Powassen....	857	Parry Sound	G. T. R.
Prairie siding	582	Kent	G. T. R.
Pratt siding....	716	Middlesex	G. T. R.
Prescott....		Grenville	
Can. Pac. station	247		C. P. R.
Grand Trunk station	309		G. T. R.
B. M., east abutment, G. T. R. bridge over C. P. R.	303·8		D. W. Com.
B. M., on G. T. R. station........	311·5		D. W. Com.
Doorstep, main entrance to custom-house	268·6		D. W. Com.
Prescott junction	302	Grenville	G. T. R.
Preston....		Waterloo	
Station....	925		G. T. R.
Ballast hill cutting, 1·9 mile, N., ground, 964; rail..	937		G. T. R.
Prince Albert....	891	Ontario	G. T. R.

ONTARIO

Locality	Elev.	County	Authority
Princeton	934	Oxford	G. T. R.
Proton		Grey	
Station	1,584		C. P. R.
Saugeen river, 2·1 miles, N., bed, 1,551; rail	1,556		C. P. R.
Puce		Essex	
Can. Pac. station	589		C. P. R.
Grand. Trunk station	586		G. T. R.
Pukatonga, Lake	1,200	Thunder Bay	C. P. R.
Pump lake	936	Algoma	Interior
Putnam	884	Middlesex	C. P. R.
Pyette hill	1,092	Grey	Admir. chart
Quay		Durham	
Station	474		G. T. R.
Ballast pit switch, 0·8 mile, N	531		G. T. R.
Queenston	349	Lincoln	M. C. R.
Rabbit lake	938	Nipissing	Geol. Surv.
Rabbit lake	940	Thunder Bay	Geol. Surv.
Rabbit mountain	1,400	Thunder Bay	Geol. Surv.
Rainy lake	748	Parry Sound	C. P. R.
Rainy lake, high water, 1,107; low water	1,099	Rainy River	C. N. R.
Rainy lake	1,438	Nipissing	C. A. R.
Rainy Lake		Nipissing	
Station	1,453		C. A. R.
Summit, 4·1 miles, W., ground, 1,536; rail	1,527		C. A. R.
Rainy River		Rainy River	
Station	1,078		C. N. R.
Rainy river, high water, 1,057; low water, 1,053; rail	1,066		C. N. R.
Raleigh	587	Kent	C. P. R.
Raleigh	1,440	Rainy River	C. P. R.
Rama	734	Ontario	G. T. R.
Ramsay		Carleton	
Station	240		N. Y. & O. R.
Summit, 1·2 mile, S	266		N. Y. & O. R.
Ramsay		Algoma	
Station (summit)	1,403		C. P. R.
Pogamasing river, 2·5 miles, E., water, 1,339; rail	1,355		C. P. R.
Ramsay lake	810	Nipissing	Geol. Surv.
Rankin lake	976	Nipissing	Geol. Surv.
Rapid lake	1,239	Nipissing	C. A. R.
Rat lake	1,084	Parry Sound	C. P. R.
Rat Portage	1,088	Rainy River	C. P. R.
Ravensworth		Parry Sound	
Station	1,414		C. A. R.
Cashman brook, 0·8 mile, E., water, 1,355; rail	1,433		C. A. R.
Tonawanda river, 2·3 miles W., bed, 1,262; rail	1,204		C. A. R.
Rawdon	449	Hastings	C. O. R.
Rayside (now Azilda)		Algoma	
Station	885		C. P. R.
Summit, 3·5 miles, E., ground, 996; rail	992		C. P. R.
Realoro	857	Victoria	G. T. R.
Red Cedar lake	900	Nipissing	Geol. Surv.
Red Deer lake	685	Nipissing	Geol. Surv.
Red Pine lake	679	Parry Sound	C. P. R.
Red Sucker	733	Thunder Bay	C. P. R.
Redwater lake, Lower	1,003	Nipissing	Geol. Surv.
Redwater lake, Upper	1,004	Nipissing	Geol. Surv.
Renfrew		Renfrew	
Can. Pac. station	416		C. P. R.
Can. Atl. station	427		C. A. R.
Bonnechère river, water, 365; rail	389		C. P. R.
Renfrew junction	404	Renfrew	C. A. R.
Renton	709	Norfolk	G. T. R.
Renwick	629	Kent	L. E. & D. R. R.
Rib or Waibikaiginaising lake	1,013	Nipissing	Geol. Surv.
Rice lake	614	Northumberland	Rys. & Canals
Richardson	600	Kent	L. E. & D. R. R.
Richmond Hill (now Maple)	817	York	G. T. R.
Richmond village	321	Carleton	T. & O. R.
Richwood	933	Oxford	G. T. R.

Locality	Elev.	County	Authority
Rideau		Frontenac	
Station ...	302		G. T. R.
B. M., course below coping, west abutment of G. T. Ry. bridge over Cataraqui river.	308·6		D. W. Com.
B. M., coping, west side of entrance lock.......	258·9		D. W. Com.
Rideau lake..........	404	Leeds	Rys. & Canals
Ridgetown		Kent	
Mich. Cent. station.	657		M. C. R.
L. E. & D. R. station	670		L. E. & D. R. R.
Ridgeville...		Welland	
Station ..	688		T. H. & B. R.
Gravel pit, 1·1 mile, N	806		T. H. & B. R.
Ridgeway....	618	Welland	G. T. R.
Ridout	1,369	Algoma	C. P. R.
Rifle Ranges ..	290	Peel	G. T. R.
Ringold...		Kent	
Station ...	582		C. P. R.
Grand Trunk crossing, 2·2 miles, E.	500		C. P. R.
Ripley....	806	Bruce	G. T. R.
Roberts (formerly Duchesnay) ...	1,377	Algoma	C. P. R.
Roberts hill...	980	Algoma	Admir. chart
Robertsville ...	753	Frontenac	K. & P. R.
Robinson lake....	953	Lanark	K. & P. R.
Rockford ...	913	Grey	C. P. R.
Rockcliffe....	478	Renfrew	C. P. R.
Rock lake....	1,281	Nipissing	C. A. R.
Rock lake....	446	Frontenac	K. & P. R.
Rock lake....	643	Algoma	B. M. & A. R.
Rock Lake station.	1,292	Nipissing	C. A. R.
Rock Lake station.	651	Algoma	B. M. & A. R.
Rockland	163	Russell	C. A. R.
Rockwood		Wellington	
Station	1,182		G. T. R.
Eramosa river, water, 1,140; rail.	1,182		G. T. R.
Summit, 1·4 mile, E.	1,218		G. T. R.
Rodney		Elgin	
Mich. Cent. station.	691		M. C. R.
L. E. & D. R. station.	694		L. E. & D. R. R.
Rolling lake....	1,536	Thunder Bay	C. N. R.
Romford	839	Nipissing	C. P. R.
Rond, Lac	1,339	Algoma	C. P. R.
Rondeau	577	Kent	L. E. & D. R. R.
Rose lake ...	1,528	Thunder Bay	Dawson
Rose Point (Parry Sound).	602	Parry Sound	C. A. R.
Rosebank	291	Ontario	G. T. R.
Rosemere ...		Thunder Bay	
Station	1,481		C. N. R.
Swamp river, 0·9 mile, W., water, 1,479; rail	1,485		C. N. R.
Ross lake....	936	Nipissing	Geol. Surv.
Rosseau, Lake	748	Muskoka	Interior
Rosslyn ...	610	Essex	M. C. R.
Rossport	643	Thunder Bay	C. P. R.
Round hill	1,118	Algoma	Admir. chart
Round or Cameron lake	628	Lanark	K. & P. R.
Round lake........	564	Renfrew	Geol. Surv.
Round lake (Whitefish river)	778	Algoma	Geol. Surv.
Round lake (Kenogami river).....	1,033	Thunder Bay	Geol. Surv.
Round lake (Gull river)......	1,375	Thunder Bay	C. P. R.
Round lake (Ravensworth).....	1,510	Nipissing	C. A. R.
Round or Mountain lake (Montreal river).....	911	Nipissing	Geol. Surv.
Roundhouse.....	598	Essex	M. C. R.
Rowan.		Thunder Bay	
Station	1,128		C. N. R.
Bay of Matawin river, 3·7 miles, W., water......	1,109		C. N. R.
Rowan, Mount.....	1,350	Thunder Bay	C. P. R.
Ruscomb.....	605	Essex	M. C. R.
Russell.....	236	Russell	N. Y. & O. R.
Russell.	415	Renfrew	C. P. R.
Ruth lake....	991	Nipissing	Geol. Surv.

7½

ONTARIO

Locality	Elev.	County	Authority
Rutherglen		Nipissing	
Station	789		C. P. R.
Summit, 2·9 miles, W	838		C. P. R.
Ruthven		Essex	
Station	692		L. E. & D. R. R.
Summit, 1·8 mile, W., ground, 734 ; rail...........	707		L. E. & D. R. R.
Rydal Bank	644		B. M. & A. R.
Rymal .... ,	644	Wentworth	G. T. R.
Nabawi (summit)..	1,480	Rainy River	C. N. R.
Sables, Lake aux	1,518	Algoma	C. P. R.
Sabourin	284	Grenville	C. P. R.
Saganaga lake	1,482	Thunder Bay	Dawson
St. Andrews lake	554	Frontenac	K. & P. R.
St. Ann	608	Lincoln	T. H. & B. R.
St. Catharines		Lincoln	
Grand Trunk station	347		G. T. R.
St. Catharines & Niagara Central station	346		St. C. & N. C. R.
St. Catharines viaduct, 0.6 mile, E., bed, 263 ; rail..	347		G. T. R.
St. Catharines and Niagara Central junction	579	Welland	M. C. R.
St. Clair station......	584	Essex	G. T. R.
St. Clair junction...	763	Elgin	M. C. R.
St. Clair, Lake.			
Standard high water..........................	577		D. W. Com.
Standard low water.	573·2		D. W. Com.
Mean water, 1871 to 1895.........	574·7		D. W. Com.
St. David..........		Welland	
Station	609		G. T. R.
St. David valley, 2·2 miles, W., ground; 480 ; rail..	517		G. T. R.
St. Eugene	181	Prescott	C. P. R.
St. George...........		Brant	
Station	725		G. T. R.
Fairchild creek (St. George viaduct), bed, 669 ; rail.	725		G. T. R.
St. George lake.....	641	Frontenac	K. & P. R.
St. Ignace, Isle.	1,863	Thunder Bay	Admir. chart
St. Jacob.	1,104	Waterloo	G. T. R.
St. Joachim....	585	Essex	C. P. R.
St. Joseph, Lake.......	1,172	Rainy River	Geol. Surv.
St. Marys		Perth	
Station	1,082		G. T. R.
Thames river, 1·0 mile, W., water, 987 ; rail......	1,040		G. T. R.
St. Marys switch....	1,058	Perth	G. T. R.
St. Ola		Hastings	
Station (summit)....	1,068		C. O. R.
Bass creek, 0·8 mile, N , water, 1,037 ; rail..	1,043		C. O. R.
St. Patrick, Mount.......	1,383	Renfrew	K. & P. R.
St. Pauls	1,165	Perth	G. T. R.
St. Thomas		Elgin	
Mich. Cent. station.	766		M. C. R.
L. E. & D. R. station	762		L. E. & D. R. R.
Grand Trunk station	752		G. T. R.
Kettle creek, 0·5 mile, W., water, 676 ; rail........	746		G. T. R.
Kettle creek, 1·4 mile, W., bed, 661 ; rail........	755		M. C. R.
Can. Pac. Ry. crossing, 1·3 mile, E.....	783		G. T. R.
City hall, floor level	754		City Eng.
Waterworks reservoir	682		City Eng.
St. Williams.............	680	Norfolk	G. T. R.
Nakwite or Dog lake........................	1,199	Rainy River	C. P. R.
Salter lake	1,478	Thunder Bay	C. P. R.
Saltfleet (now Vinemount)	648	Wentworth	T. H. & B. R.
Salvation lake	1,008	Nipissing	Geol. Surv.
Sand lake	383	Leeds	Rys. & Canals
Sand or Wazuzke lake........................	1,111	Parry Sound	Geol. Surv.
Sand lake·	1,530	Thunder Bay	Geol. Surv.
Sandbar lake	1,035	Rainy River	Geol. Surv.
Sandison.	648	Kent	L. E. & D. R. R.
Sand Point.	266	Renfrew	C. P. R.
Sand Point lake	1,337	Rainy River	C. P. R.
Sandy lake	664	Parry Sound	C. P. R.
Sandy lake	1,190	Rainy River	C. P. R.

ONTARIO

Locality	Elev.	County	Authority
Sarn...		Lambton	
Grand Trunk station	587		G.T.R.
L.E. & D.R. station	600		L.E. & D.R.R.
Sarnia tunnel—			
Canadian portal, rail.	547		G.T.R.
Lowest point, rail	508		G.T.R.
American portal, rail.	557		G.T.R.
St. Clair river—			
Standard low water	578·3		D.W.Com.
Standard high water.	583·1		D.W.Com.
Mean water (1871-1900)	579·7		D.W.Com.
Sarnia Tunnel station.	612	Lambton	G.T.R.
Sault Ste. Marie.		Algoma	
Can. Pac. station.	632		C.P.R.
Algoma Cent. station.	585		A.C. & H.B.R.
Canadian lock, lower sill, 561·5; upper sill, 576·6; coping.	605·4		D.W.Com.
Sault canal, lower entrance—			
Standard low water	581·8*		D.W.Com.
Standard high water	586·8*		D.W.Com.
Mean water (1871-1895)	583·4*		D.W.Com.
Sault canal, upper entrance—			
Standard low water	600·1		D.W.Com.
Standard high water	604·3		D.W.Com.
Mean water (1871-1895)	601·0		D.W.Com.
Sanrin.	701	Simcoe	G.T.R.
Savanne	1,506	Thunder Bay	C.P.R.
Savanne lake or lac du Milieu.	1,536	Thunder Bay	Dawson
Sayre lake.	1,222	Algoma	A.C. & H.B.R.
Scanlon.	742	Simcoe	G.T.R.
Scarborough		York	
Station.	546		G.T.R.
Summit, 0·5 mile, E., ground, 564; rail	557		G.T.R.
Schaw.	985	Wellington	C.P.R.
Schepeler.	730	Ontario	G.T.R.
Schreiber		Thunder Bay	
Station.	983		C.P.R.
Summit, 1·4 mile, W., ground, 1,037; rail.	1,023		C.P.R.
Schreiber hills	1,450	Thunder Bay	Admir. chart
Scotia.		Parry Sound	
Station.	1,088		G.T.R.
Summit, 1·7 mile, S., ground, 1,151; rail.	1,138		G.T.R.
Scotland.	780	Brant	T.H. & B.R.
Scovil.	1,187	Rainy River	C.P.R.
Scugog, Lake.	821	Durham	Rys. & Canals
Sea Cliff Park.	577	Essex	M.C.R.
Seaforth.	1,008	Huron	G.T.R.
Seagrave.	839	Ontario	G.T.R.
Sebringville	1,172	Perth	G.T.R.
Seeley.	341	Leeds	B. & W.R.
Seguin Falls	963	Parry Sound	C.A.R.
Sesebe lake.	922	Parry Sound	Geol. Surv.
Seul, Lac	1,140	Rainy River	Geol. Surv.
Seven-league lake, high water, 541; low water	522	Nipissing	Rys. & Canals
Severn		Muskoka	
Station	730		G.T.R.
Severn river, water, 697; rail.	724		G.T.R.
Shabakwa.		Thunder Bay	
Station	1,241		C.N.R.
Matawin river, 3·3 miles, E., mouth of Shebandowan river, water, 1,203; rail	1,217		C.N.R.
Shebandowan river, 0·9 mile, W., water, 1,284; rail	1,297		C.N.R.
Shakespeare.	1,182	Perth	G.T.R.
Shallow Lake station.	739	Grey	G.T.R.

* These figures—corrected in accordance with the latest determination of the level of L. Superior—are extracted from the report of the U.S. Deep Waterways Commission for 1895, but the Land Commissioner of the Algoma Central & Hudson Bay Railway states that the recent improvements in the Hay Lake channel and at the Neebish rapids have increased the "head" of water in the Sault power canals fully 2 feet which indicates a corresponding fall in the level of the water below the lock.

ONTARIO

Locality	Elev.	County	Authority
Shannonville....		Hastings	
Station	334		G.T.R.
Salmon river, 1·8 mile, E. water, 268; rail	283		G.T.R.
Sharbot Lake station	646	Frontenac	C.P.R.
Sharbot lake	639	Frontenac	C.P.R.
Sharp lake	905	Nipissing	Geol. Surv.
Shaw lake	728	Parry Sound	C.P.R.
Shebandowan	1,540	Thunder Bay	C.P.R.
Shebandowan lake, Lower	1,475	Thunder Bay	C.N.R.
Shedden		Elgin	
Mich. Cent. station	725		M.C.R.
L.E. & D.R. station	726		L.E. & D.R.R.
Sheguiandah hill	1,015	Algoma	Admir. chart
Shelburne	1,625	Dufferin	C.P.R.
Shell or Esnagami lake	1,273	Algoma	C.P.R.
Sherk	618	Welland	G.T.R.
Shrewsbury	586	Kent	L.E. & D.R.R.
Sidney	301	Hastings	G.T.R.
Sigsworth	511	Frontenac	K. & P.R.
Silver lake (Oso)	578	Frontenac	K. & P.R.
Silver lake (Portland)	446	Frontenac	K. & P.R.
Silver Creek station	807	Thunder Bay	C.N.R.
Silver Creek		Simcoe	
Station	744		G.T.R.
Summit, 0·8 mile, S	776		G.T.R.
Silverdale	611	Lincoln	T.H. & B.R.
Silver Mountain station	1,282	Thunder Bay	C.N.R.
Silver mountain	1,730	Thunder Bay	Geol. Surv.
Simcoe		Norfolk	
Loop line station	718		G.T.R.
G.B. & L.E. station	712		G.T.R.
Simcoe, Lake, flood (1871), 728; low water (1900)	719	Simcoe	G.T.R.
Simpson lake	955	Nipissing	Geol. Surv.
Singleton lake	271	Leeds	Rys. & Canals
Skead	214	Carleton	C.P.R.
Skinner bluff	951	Grey	Admir. chart
Smith, C···	950	Algoma	Admir. chart
Smiths ...ns.		Lanark	
Station	423		C.P.R.
Rideau canal, 1·3 mile, S., water, 348; rail	380		C.P.R.
Smithville	625	Lincoln	T.H. & B.R.
Smoky head	752	Bruce	Admir. chart
Snake Island lake	953	Nipissing	Geol. Surv.
Snake River		Renfrew	
Station	424		C.P.R.
Snake river, water, 406; rail	416		C.P.R.
Snapper Bridge	277	Leeds	T.I.R.
Snedden	379	Lanark	C.P.R.
Snelgrove	838	Peel	C.P.R.
Snow Road	701	Frontenac	K. & P.R.
Sombra	588	Lambton	L.E. & D.R.R.
Sonya	901	Ontario	G.T.R.
Soperton	365	Leeds	B. & W.R.
Soulier lake	1,023	Algoma	A.C. & H.B.R.
Source lake	1,467	Nipissing	C.A.R.
South lake	1,496	Rainy River	C.P.R.
South lake	1,551	Thunder Bay	Dawson
Southampton	614	Bruce	G.T.R.
South Indian	232	Russell	C.A.R.
South March	285	Carleton	C.A.R.
South Parkdale	285	York	G.T.R.
South River		Parry Sound	
Station	1,161		G.T.R.
South river, water, 1,147; rail	1,158		G.T.R.
Southwold	746	Elgin	M.C.R.
Spanish	608	Algoma	C.P.R.
Spanish Forks (now Forks)		Algoma	
Station	1,203		C.P.R.
Bay of Spanish river, 0·5 mile, N., water, 1,189; rail	1,201		C.P.R.

ONTARIO

Locality	Elev.	County	Authority
Spencerville..		Grenville	
Station	315		C. P. R.
South Nation river, 0·6 mile, S., water, 290; rail..	306		C. P. R.
Sphene lake	1,160	Rainy River	C. P. R.
Spiers		Wellington	
Station	1,391		C. P. R.
Grand river, 1·4 mile, E., bed, 1,350; rail	1,366		C. P. R.
Spragge (formerly Cook Mills)	599	Algoma	C. P. R.
Springbrook	533	Hastings	C. O. R.
Springfield (now Erindale)	474	Peel	C. P. R.
Springfield	793	Elgin	M. C. R.
Springford.	823	Oxford	G. T. R.
Spruce lake	978	Nipissing	Geol. Surv.
Sprucedale	1,075	Parry Sound	C. A. R.
Sprucewood (formerly Trout Creek)	680	Thunder Bay	C. P. R.
Stafford	474	Renfrew	C. P. R.
Staffordville	749	Elgin	T. L. E. & P. R.
Stamford		Welland	
Grand Trunk station	643		G. T. R.
Mich. Cent. station	618		M. C. R.
Stanawan or Grassy River lake	1,206	Rainy River	C. P. R.
Stanley		Thunder Bay	
Station	722		C. N. R.
Kaministikwia river, 1·3 mile, S., water, 680; rail.	717		C. N. R.
Stanley	688	Algoma	C. P. R.
Staples	613	Essex	M. C. R.
Stardale (summit)	291	Prescott	C. P. R.
Stayner	716	Simcoe	G. T. R.
Steel	741	Thunder Bay	C. P. R.
Steel lake	695	Thunder Bay	C. P. R.
Steeprock	1,279	Rainy River	C. N. R.
Steep Rock lake	1,207	Rainy River	C. N. R.
Stevensville.		Welland	
Grand Trunk station	585		G. T. R.
Mich. Cent. station	587		M. C. R.
Stewart lake	1,303	Rainy River	C. P. R.
Stewarton		Halton	
Station	873		G. T. R.
Credit mill-pond	822		G. T. R.
Stinson	862	Nipissing	C. P. R.
Stirling		Hastings	
Station	405		G. T. R.
Central Ontario Railway crossing, 2·2 miles, W.	395		G. T. R.
Stittsville		Carleton	
Station	386		C. P. R.
Summit, 3·7 miles, W.	487		C. P. R.
Stobie	592	Algoma	C. P. R.
Stobie mine, end of track	956	Nipissing	C. P. R.
Stoco lake.	458	Hastings	C. P. R.
Stoco	470	Hastings	B. of Q. R.
Stoney Creek		Wentworth	
Grand Trunk station	274		G. T. R.
T. H. & B. station	432		T. H. & B. R.
Stoney Point	584	Essex	G. T. R.
Stony lake	768	Peterborough	Rys. & Canals
Stony lake	1,085	Parry Sound	G. T. R.
Story	358	Leeds	C. P. R.
Stouffville	892	York	G. T. R.
Stralak (formerly Straight Lake)	1,373	Algoma	C. P. R.
Stratford		Perth	
Station	1,189		G. T. R.
Stratford & Port Dover branch crossing..	1,193		G. T. R.
Avon river (main line), water, 1166; rail	1,183		G. T. R.
Avon river (B. & G. branch), water, 1138; rail	1,171		G. T. R.
B. M., point on ledge, S.E. angle of house, N.E. corner Queen & Douro streets	1,203		City Eng.
B. M., north end of Grand Trunk Ry., B. & G. branch, arch abutment, north of St. David st..	1,178		City Eng.
Top of hydrant, corner of Ontario and Waterloo sts.	1,214		City Eng.

ONTARIO

Locality	Elev.	County	Authority
Stratford		Perth	
St. James cathedral, N.W. angle of door-sill, west door	1,208		City Eng.
City hall, floor level	1,209		City Eng.
Strathroy		Middlesex	
Station	744		G. T. R.
Sydenham river, 0 6 mile, W., water, 729; rail	730		G. T. R.
Stratton	1,113	Rainy River	C. N. R.
Streetsville		Peel	
Station	500		C. P. R.
Credit river, 0 9 mile, E., bed, 452; rail	433		C. P. R.
Streetsville junction	549	Peel	C. P. R.
Stroanness	584	Haldimand	G. T. R.
Sturgeon lake	812	Victoria	Rys. & Canals
Sturgeon lake (Kaministikwia canoe-route)	1,220	Rainy River	Dawson
Sturgeon lake (English river)	1,327	Rainy River	C. P. R.
Sturgeon Bay station	590	Simcoe	G. T. R.
Sturgeon Falls		Nipissing	
Station	685		C. P. R.
Sturgeon river, bed, 636; rail	684		C. P. R.
Sucker lake	450	Frontenac	K. & P. R.
Sudbury	850	Nipissing	C. P. R.
Suffel	264	Dundas	C. P. R.
Summerstown		Glengarry	
Station	183		G. T. R.
B.M., west wall of culvert 2 2 miles, E	169		D. W. Com.
Summit		Wentworth	
Station	780		T. H. & B. R.
Summit, 0 5 mile, E.	796		T. H. & B. R.
Summit		Rainy River	
Station	1,347		C. P. R.
Summit, 1 7 mile, E., ground, 1,303; rail	1,396		C. P. R.
Summit lake (Summit station)	1,394	Rainy River	C. P. R.
Summit lake (Kalmar)	1,253	Rainy River	C. P. R.
Summit lake	1,164	Nipissing	Geol. Surv.
Sunderland	862	Ontario	G. T. R.
Sundridge		Parry Sound	
Station	1,102		G. T. R.
Summit, 2 6 miles, N., ground, 1,195; rail	1,188		G. T. R.
Sunnidale	724	Simcoe	G. T. R.
Sunshine		Thunder Bay	
Station	1,053		C. P. R.
Matawin river, low water, 1,082; high water, 1,089; rail	1,098		C. P. R.
Superior, Lake			
Standard high water (assumed high water of 1838)	604 8		D. W. Com.
Standard low water	600 8		D. W. Com.
Extreme low water, (March, 1854)	598 1		D. W. Com.
Mean water, (1860 to 1900)	601 8		D. W. Com.
Mean water, (1871 to 1900)	601 7		D. W. Com.
Surprise lake	714	Parry Sound	C. P. R.
Sutton	758	York	G. T. R.
Swansea	254	York	G. T. R.
Sydenham	433	Frontenac	B. of Q. R.
Sydenham lake, high water, 425; low water	422	Frontenac	K. & P. R.
Taché		Rainy River	
Station	1,366		C. P. R.
Little Wabigoon river, bed, 1,348; rail	1,366		C. P. R.
Tait	640	Simcoe	G. T. R.
Talon lake, high water, 639; low water	634	Nipissing	Rys. & Canals
Tamarac		Rainy River	
Station	1,513		C. P. R.
Summit, 1 2 mile, W.	1,550		C. P. R.
Tamworth	515	Lennox & Add'ton	B. of Q. R.
Tanner lake	1,199	Rainy River	Dawson
Tansley (formerly St. Ann)		Halton	
Station	516		G. T. R.
Twelve-mile creek, 0 5 mile, N., water, 434; rail	523		G. T. R.
Tattnall lake	1,244	Thunder Bay	C. P. R.

ONTARIO

Locality	Elev.	County	Authority
Tavistock.....		Perth	
B. & G. station....	1,134		G. T. R.
G. B. & L. E. station..	1,143		G. T. R.
Tavistock junction....	1,162	Perth	G. T. R.
Taylor....	716	Elgin	M. C. R.
Tecumseh....	590	Essex	G. T. R.
Teeswater..	1,019	Bruce	C. P. R.
Temagami lake....	964	Nipissing	C. P. R.
Terra Cotta....	806	Peel	G. T. R.
Tetu, Lake....	991	Rainy River	Dawson
Thamesford....	951	Oxford	C. P. R.
Thames River station....	717	Middlesex	G. T. R.
Thamesville....		Kent	
Station....	621		G. T. R.
Thames River, 1·0 mile W., bed, 587; rail....	621		G. T. R.
The Brook station....	216	Russell	C. P. R.
'The Caves' bluff....	1,625	Grey	Admir. chart
Thedford....	681	Lambton	G. T. R.
'The Notch'....	1,102	Algoma	Admir. chart
'The Rock,' Manitowaning....	746	Algoma	Admir. chart
Thessalon....	599	Algoma	C. P. R.
Thieving-bear lake....	975	Nipissing	Geol. Surv.
Thirteen-island lake....	497	Frontenac	K. & P. R.
Thirty-island lake....	507	Frontenac	K. & P. R.
Thirty-three-island lake....	449	Frontenac	K. & P. R.
T.......		Renfrew	
station....	510		C. P. R.
Summit, 2·0 miles, S....	517		C. P. R.
Thompson Mills..	363	Lennox & Add'ton	B. of Q. R.
Thompsonville....		Simcoe	
Station....	723		G. T. R.
Nottawasaga river, water, 700; rail....	719		G. T. R.
Thornbury....	612	Grey	G. T. R.
Thorncliff....	609	Nipissing	C. P. R.
Thorndale....	834	Middlesex	G. T. R.
Thornhill....	636	York	G. T. R.
Thornton....	943	Simcoe	G. T. R.
Thorold....		Welland	
Grand Trunk station....	526		G. T. R.
St. C. & N. C. station....	595		St. C. & N. C. R.
Three Portage lake....	966	Nipissing	Geol. Surv.
Thunder cape....	1,950	Thunder Bay	Admir. chart
Tilbury....	582	Essex	C. P. R.
Tilbury....	590	Kent	M. C. R.
Tilden lake....	928	Nipissing	Geol. Surv.
Tillsonburg....		Oxford	
Grand Trunk station....	757		G. T. R.
B. & T. station....	761		G. T. R.
Mich. Cent. station....	804		M. C. R.
Grand Trunk Ry. crossing, G. T. Ry. rail, 782;			
M. C. R. R., rail....	863		M.C. R.
Otter creek, high water, 662; low water, 653; rail..	760		G. T. R.
Tillsonburg junction....		Oxford	
Station....	799		G. T. R.
Summit, 0·4 mile, W., ground, 820; rail....	810		G. T. R.
Timiskaming, Lake, high water, 592; low water....	578	Nipissing	Rys. & Canals
Tioga....		Simcoe	
Station....	754		G. T. R.
Pine river, water, 720; rail....	754		G. T. R.
Tip-top hill..	2,120	Thunder Bay	Admir. chart
Tomiko lake....	795	Nipissing	Geol. Surv.
Toronto....		York	
Union station....	254		G. T. R.
Queen street....	260		C. P. R.
Winchester street....	266		C. P. R.
Leaside junction..	429		C. P. R.
North Toronto station....	400		C. P. R.
Parkdale....	305		C. P. R.
Bloor street....	371		C. P. R.

ONTARIO

Locality	Elev.	County	Authority
Toronto		York	
Don station	252		G. T. R.
North Parkdale	303		G. T. R.
South Parkdale	285		G. T. R.
Bloor street	369		G. T. R.
Carlton	405		G. T. R.
Scott street	256		G. T. R.
Dufferin street	292		G. T. R.
Queen St. East	293		G. T. R.
Winchester street	269		G. T. R.
Don Valley	262		G. T. R.
Rosedale	298		G. T. R.
Moore Park	451		G. T. R.
Yonge St. North	505		G. T. R.
Upper Canada College	510		G. T. R.
Eglinton Avenue	537		G. T. R.
Forest Hill	577		G. T. R.
Fairbank station	580		G. T. R.
Fairbank junction	511		G. T. R.
Symes	383		G. T. R.
Lambton	389		G. T. R.
Bloor St. West	322		G. T. R.
City hall, ground floor	296		City Eng.
City hall, top of tower	584		City Eng.
Reservoir, water at zero	420		City Eng.
Corner of King and Yonge streets	273		City Eng.
Parliament buildings, ground floor	356		City Eng.
Parliament buildings, roof	648		City Eng.
Woodlawn avenue, highest point in city	469		City Eng.
City datum	204·8		City Eng.
Harbour Commissioner's gauge, zero	244·8		Interior
Toronto junction	394	York	C. P. R.
Tory Hill station	1,147	Haliburton	I. B. & O. R.
Tottenham	834	Simcoe	G. T. R.
Towner	1,035	Parry Sound	G. T. R.
Townsend	737	Norfolk	M. C. R.
Travers lake	625	Nipissing	Geol. Surv.
Trembly	1,074	Algoma	A. C. & H. B. R.
Trenton		Northumberland	
Grand Trunk station	280		G. T. R.
Cent. Ont. station	258		C. O. R.
Trout lake, high water, 1,246; low water	1,243	Renfrew	K. & P. R.
Trout lake (Mattawa river) high water, 668; low water.	665	Nipissing	Rys. & Canals
Trout lake (Petawawa river)	863	Nipissing	Geol. Surv.
Trout lake	857	Nipissing	Geol. Surv.
Trout lake, Upper	944	Frontenac	K. & P. R.
Trout lake, Lower	925	Frontenac	K. & P. R.
Trout Creek station	1,034	Parry Sound	G. T. R.
Trudeau		Thunder Bay	
Station	1,042		C. P. R.
Summit, 3·0 miles, E., ground, 1,151; rail	1,124		C. P. R.
Tupperville	604	Kent	L. E. & D. R. R.
Turnbull		Algoma	
Station	1,390		C. P. R.
Pogamasing river, water, 1,383; rail	1,392		C. P. R.
Turner	744	Bruce	G. T. R.
Turner lake	1,057	Nipissing	Geol. Surv.
Turtle lake, high water, 666; low water	664	Nipissing	Rys. & Canals
Turtle lake (Turtle river)	1,133	Rainy River	C. P. F
Turtle lake (Jackpine)	1,366	Rainy River	C. P. R.
Tweed	476	Hastings	C. P. R.
Twelfth Line	676	Lambton	G. T. R.
Twin lake	1,484	Thunder Bay	Crown Lands
Twin lake, Lower	977	Nipissing	Geol. Surv.
Twin lake, Upper	993	Nipissing	Geol. Surv.
Two Rivers, Lake of	1,289	Nipissing	C. A. R.
Tyndall		Haliburton	
Station	1,072		I. B. & O. R.
Burnt river, 1·0 mile, W., bed, 1,033; rail	1,050		I. B. & O. R.
Uhtoff	697	Simcoe	G. T. R.

ONTARIO

Locality	Elev.	County	Authority
Ulverston lake	1,358	Rainy River	C. P. R.
Unionville		York	
Station	575		G. T. R.
Rouge river, water, 551 ; rail	575		G. T. R.
Upper Rideau lake	408	Leeds	Rys. & Canals
Upsala	1,562	Thunder Bay	C. P. R.
Uptergrove	736	Ontario	G. T. R.
Utopia		Simcoe	
Station	693		G. T. R.
Nottawasaga river, bed, 581 ; rail	630		G. T. R.
Utterson		Muskoka	
Station	1,041		G. T. R.
Summit, 0·7 mile, N., ground, 1,071 ; rail	1,057		G. T. R.
Uxbridge	886	Ontario	G. T. R.
Vanessa	783	Norfolk	T. H. & B. R.
Vankleek Hill		Prescott	
Can. Atl. station	297		C. A. R.
Can. Pac. station	272		C. P. R.
Summit, 1·4 mile, S	322		C. A. R.
Can. Pac. Ry. crossing	263		C. A. R.
Varty lake	463	Lennox & Add'ton	B. of Q. R.
Vermilion Bay station	1,221	Rainy River	C. P. R.
Vermilion or Mazininwaning lake	786	Algoma	Geol. Surv.
Verner	669	Nipissing	C. P. R.
Vernon lake	936	Muskoka	G. T. R.
Verona		Frontenac	
Station	459		K. & P. R.
Napanee river, 1·5 mile, S., water, 448 ; rail	456		K. & P. R.
Vespra	729	Simcoe	G. T. R.
Viaduct lake	1,248	Rainy River	C. P. R.
Victoria	605	Welland	M. C. R.
Victoria, Mount	1,062	Algoma	Admir. chart
Victoria P. O	850	Thunder Bay	Geol. Surv.
Victoria Harbour	595	Simcoe	G. T. R.
Victoria junction	862	Victoria	G. T. R.
Victoria Mine	822	Algoma	C. P. R.
Victoria Road	847	Victoria	G. T. R.
Vienna	650	Elgin	T. L. E. & P. R.
Villa Nova	730	Norfolk	M. C. R.
Vine	924	Simcoe	G. T. R.
Vinemount	648	Wentworth	T. H. & B. R.
Vittoria	679	Norfolk	G. T. R.
Vivian	990	York	G. T. R.
Vosburg	608	Kent	G. T. R.
Wabaunk lake	841	Nipissing	Geol. Surv.
Wabigoon station	1,211	Rainy River	C. P. R.
Wabigoon lake, Little	1,206	Rainy River	C. P. R.
Wabigoon lake	1,206	Rainy River	C. P. R.
Wabiteek lake	1,483	Thunder Bay	C. P. R.
Wagong or Cypress lake	1,428	Algoma	C. P. R.
Wakemika lake	935	Nipissing	Geol. Surv.
Waibikaiginaising or Rib lake	1,013	Nipissing	Geol. Surv.
Wainfleet	590	Welland	G. T. R.
Waldemar		Dufferin	
Station	1,487		C. P. R.
Grand river, bed, 1,454 ; rail	1,480		C. P. R.
Wales		Stormont	
Station	235		G. T. R.
B. M., door-sill, front of station	238·8		D. W. Com.
Walford	608	Algoma	C. P. R.
Walker	731	Middlesex	M. C. R.
Walkerton	931	Bruce	G. T. R.
Walkerville		Essex	
Can. Pac. station	615		C. P. R.
Grand Trunk station	588		G. T. R.
Wallaceburg		Kent	
Station	583		L. E. & D. R. R.
Sydenham river, 0·7 mile, E., water, 576 ; rail	585		L. E. & D. R. R.

ONTARIO

Locality	Elev.	County	Authority
Walsh		Norfolk	
Station	738		G. T. R.
Summit, 0·2 mile, N., ground, 754; rail.·	739		G. T. R.
Wanapitai.............		Nipissing	
Station	779		C. P. R.
Wanapitai river, high water, 772; rail..	779		C. P. R.
Wanapitai lake, ice, 851; extreme high water.... ...	862	Nipissing	Geol. Surv.
Wangoon, Lake, high water.	1,486	Algoma	C. P. R.
Wanstead	700	Lambton	G. T. R.
War Eagle lake..........................	1,080	Rainy River	C. P. R.
Warkworth		Northumberland	
Village	470		C. N. & P. R.
Mill-pond.	468		C. N. & P. R.
Warren........................	689	Nipissing	C. P. R.
Wasacsinagama lake..	1,025	Nipissing	Geol. Surv.
Washago	728	Simcoe	G. T. R.
Washaiganog or Fairbank lake..............	867	Algoma	Geol. Surv.
Watap or Mud lake...................	1,667	Thunder Bay	Dawson
Waterdown	341	Wentworth	G. T. R.
Waterford	761	Norfolk	M. C. R.
Waterloo..........................	1,056	Waterloo	G. T. R.
Watford..		Lambton	
Station	783		G. T. R.
Brown creek, water, 761; rail..... ....,	792		G. T. R.
Summit, 3·3 miles, E. .... ,,,.........	813		G. T. R.
Watson.....	602	Lambton	L. E. & D. R. R.
Waubaushene	593	Simcoe	G. T. R.
Waubuno	867	Middlesex	G. T. R.
Wawa.....	1,060	Algoma	A. C. & H. B. R.
Wawa lake	938	Algoma	A. C. & H. B. R.
Wawanosh....	601	Lambton	L. E. & D. R. R.
Wawaskesh lake	720	Parry Sound	Geol. Surv.
Wawiagama lake	917	Nipissing	Geol. Surv.
Wazuzke or Sand lake	1,111	Parry Sound	Geol. Surv.
Webbwood.	661	Algoma	C. P. R.
Weidman	682	Lambton	M. C. R.
Wekwemikong hill	905	Algoma	Admir. chart
Welland		Welland	
Grand Trunk station..... .· ....:..	691		G. T. R.
Mich, Cent. station..................	585		M. C. R.
Welland junction....................	577	Welland	G. T. R.
Weller Bay station..................	262	Prince Edward	C. O. R.
Wellington....		Prince Edward	
Station	304		C. O. R.
Summit, 1·7 mile, W.	337		C. O. R.
Welsh	431	Lanark	C. P. R.
Wenebegon lake.	1,450	Algoma	C. P. R.
West knob..	1,164	Algoma	Admir. chart
West Brantford......		Brant	
Grand Trunk station..	667		G. T. R.
T. H. & B. station	664		T. H. & B. R.
West Lorne	712	Elgin	L. E. & D. R. R.
Westminster........................	878	Middlesex	L. E. & D. R. R.
Weston		York	
Grand Trunk station....	425		G. T. R.
Can. Pac. station....... ...	427		C. P. R.
Humber river, 1·1 mile, W., water, 394; rail.	458		G. T. R.
Westport	455	Leeds	B. & W. R.
Wexford.......................	541	York	C. P. R.
Wheatley......................	609	Essex	L. E. & D. R. R.
Whitby......................	276	Ontario	G. T. R.
White	727	Elgin	L. E. & D. R. R.
White lake	1,073	Thunder Bay	C. P. R.
White lake.....	689	Frontenac	K. & P. R.
White lake.....	540	Lennox & Add'ton	B. of Q. R.
White mountain....	1,335	Algoma	Admir. chart
White Bear lake..	942	Nipissing	Geol. Surv.

ONTARIO

Locality	Elev.	County	Authority
White Beaver lake	841	Nipissing	Geol. Surv.
Whitebread..............	585	Kent	L. E. & D. R. R.
Whitechurch.................	1,046	Bruce	G. T. R.
Whitefish.........		Algoma	
Station..... ..	808		C. P. R.
Vermilion river, 2·2 miles, E., water, 760 ; rail.....	786		C. P. R.
Whitefish Lake		Thunder Bay	
Station...	1,348		C. N. R.
Summit, 9·4 miles, W........	1,552		C. N. R.
Whitefish lake (Silver mountain)	1,330	Thunder Bay	C. N. R.
Whitefish lake (Michipicoten river)	986	Thunder Bay	C. P. R.
Whitefish lake (E. of Nipigon).............	1,244	Thunder Bay	C. P. R.
Whitefish lake..	323	Leeds	Rys. & Canals
Whitefish lake..	1,281	Nipissing	C. A. R.
Whitefish lake..	1,214	Rainy River	C. P. R.
Whitehall	1,097	Parry Sound	C. A. R.
White Loon lake....·.............	739	Frontenac	K. & P. R.
White River..		Algoma	
Station..............	1,230		C. P. R.
Summit, 4·2 miles, W., ground, 1,295 ; rail........	1,269		C. P. R.
White Rock lake.......	1,230	Thunder Bay	C. P. R.
White Trout lake	1,271	Nipissing	Geol. Surv.
Whitewater lake.	963	Algoma	C. P. R.
Whitewater or Nitawagami lake.....	835	Algoma	Geol. Surv.
Whitney...		Nipissing	
Station.............	1,262		C. A. R
Summit, 3·0 miles, W., ground, 1,330 ; rail........	1,320		C. A. R.
Whitson lake.....	938	Nipissing	Geol. Surv.
Wick........	865	Ontario	G. T. R.
Wicksteed lake..............	941	Nipissing	Geol. Surv.
Wigle........	640	Essex	M. C. R.
Wigwasikagamog lake..............	682	Nipissing	C. P. R.
Wilberforce..............		Haliburton	
Station.,..	1,194		I. B. & O. R.
Burnt river, water, 1,184 ; rail......	1,194		I. B. & O. R.
Wilbur		Frontenac	
Station.............	894		K. & P. R.
Summit, 0·4 mile, N	905		K. & P. R.
Wilbur mines..............	895	Lanark	K. & P. R.
Wilbur Mines junction..............	840	Frontenac	K. & P. R.
Wilcox lake..............	1,030	Rainy River	Geol. Surv.
Wilkie	664	Kent	L. E. & D. R. R.
Williams..............		Algoma	
Station...	1,190		C. P. R.
Magpie river and lake, 3·3 miles, W., water, 1,138 ; rail	1,164		C. P. R.
Wilno		Renfrew	
Station....	961		C. A. R.
Summit of Hagarty pass, 2·0 miles, W........ ..	1,026		C. A. R.
Wilson........	555	Lennox & Add'ton	B. of Q. R.
Wilson lake..............	1,173	Nipissing	Geol. Surv.
Winchester..............	247	Dundas	C. P. R.
Windermere..............		Algoma	
Station..............	1,437		C. P. R.
Summit, 1·7 mile, W., ground, 1,475 ; rail.....	1,463		C. P. R.
Windermere lake, high water, 1,413 ; low water........	1,408	Algoma	C. P. R.
Windham..............	815	Norfolk	M. C. R.
Windigo..............	1,486	Thunder Bay	C. N. R.
Windigustigwan lake..............	1,443	Thunder Bay	C. N. R.
Windmill Point.......	599	Welland	G. T. R.
Windsor.		Essex	
Can. Pac. station.	581		C. P. R.
Grand Trunk station......	579		G. T. R.
Mich. Cent. station.............	580		M. C. R.
Detroit river at Windsor, high water, 577·3 ; low water, 573·5 ; mean level..............	575·0		D. W. Com.
Windy Lake station..............	1,221	Algoma	C. P. R.
Windy or Makoping lake..............	1,060	Algoma	Geol. Surv.

ONTARIO

Locality	Elev.	County	Authority
Wingham		Huron	
Can. Pac. station	1,020		C. P. R.
Grand Trunk station.	1,044		G. T. R.
Maitland river, high water, 1,012 ; low water, 1,007 ;			
rail (Kincardine branch).............	1,042		G. T. R.
Wingham junction..............	1,078	Huron	G. T. R.
Winjiget lake............	1,265	Algoma	C. P. R.
Winnebago		Algoma	
Station	1,461		C. P. R.
Winnebago river, 0·9 mile, W., water, 1,429 ; rail..	1,447		C. P. R.
Summit, 1·2 mile, E.............	1,499		C. P. R.
Winona	285	Wentworth	G. T. R.
Winston station	762	Thunder Bay	C. P. R.
Winston hill..............	1,600	Thunder Bay	Admir. chart
Wisner	719	Brant	T. H. & B. R.
Woito....		Renfrew	
Station	652		C. A. R.
Summit, 0·3 mile, N., ground, 660 ; rail.....	653		C. A. R.
Wolf	607	Thunder Bay	C. P. R.
Wolford........	396	Grenville	C. P. R.
Wolverton	958	Oxford	C. P. R.
Woman River....		Algoma	
Station....	1,444		C. P. R.
Woman river, water, 1,437 ; rail....	1,444		C. P. R.
Summit, 2·4 miles, W., ground, 1,529 ; rail...	1,516		C. P. R.
Woodbridge..	554	York	C. P. R.
Woods, Lake of the..		Rainy River	
High water (1879).............	1,062		C. P. R.
Low water....	1,054		C. P. R.
Mean water...	1,057		C. P. R.
Woodslee....	615	Essex	M. C. R.
Woodstock....		Oxford	
Grand Trunk station.	958		G. T. R.
Can. Pac. Ry. station	947		C. P. R.
Can. Pac. Ry. Ingersoll branch, crossing......	920		G. T. R.
Grand Trunk Ry., Stratford and Port Dover branch,			
crossing	947		C. P. R.
Thames river, bed, 910 ; rail............	932		C. P. R.
Summit, 2·9 miles, E., ground, 1,022 ; rail.	1,007		G. T. R.
Woodville	906	Victoria	G. T. R.
Woonga	1,518	Rainy River	C. P. R.
Worthington	775	Algoma	C. P. R.
Wroxeter	1,104	Huron	C. P. R.
Wyevale.....	772	Simcoe	G. T. R.
Wylie.......	528	Renfrew	C. P. R.
Wyoming............	709	Lambton	G. T. R.
Wyton..........	907	Middlesex	G. T. R.
Yarker		Lennox & Add'ton	
Station	451		B. of Q. R.
Napanee river, bed, 431; rail..	462		B. of Q. R.
Yarmouth....	769	Elgin	M. C. R.
Yarmouth............	782	Elgin	L. E & D. R. R.
York	662	Haldimand	G. T. R.
York....	425	York	G. T. R.
York Road	461	York	G. T. R.
Zealand.....		Frontenac	
Station	689		C. P. R.
Summit, 1·4 mile, W........	732		C. P. R.
Zephyr	767	York	G. T. R.

PRINCE EDWARD ISLAND

Locality	Elev.	County	Authority
Albany	127	Prince	P. E. I. R.
Alberton		Prince	
Station	24		P. E. I. R.
Wharf	6		P. E. I. R.
Alma (Montrose)	142	Prince	P. E. I. R.
Ashton	143	Kings	P. E. I. R.
Baldwin		Kings	
Station	189		P. E. I. R.
Summit, 0·1 mile, S.	195		P. E. I. R.
Barbara Weit	30	Prince	P. E. I. R.
Bear River		Kings	
Station	219		P. E. I. R.
Summit, 0·2 mile, E.	228		P. E. I. R.
Bedford	92	Queens	P. E. I. R.
Breadalbane	135	Queens	P. E. I. R.
Bloomfield	130	Prince	P. E. I. R.
Blooshanks	124	Prince	P. E. I. R.
Brudenell	65	Kings	P. E. I. R.
Cape Traverse	11	Prince	P. E. I. R.
Cardigan	79	Kings	P. E. I. R.
Cemetery	127	Queens	P. E. I. R.
Charlottetown		Queens	
Station	8		P. E. I. R.
Intercol. bridge, rail	17		P. E. I. R.
Clyde	202	Queens	P. E. I. R.
Coleman	67	Prince	P. E. I. R.
Colville	179	Queens	P. E. I. R.
Conway	73	Prince	P. E. I. R.
De-Blois	124	Prince	P. E. I. R.
Douglas	88	Kings	P. E. I. R.
Dundee	108	Kings	P. E. I. R.
Ellerslie	58	Prince	P. E. I. R.
Elliott	181	Queens	P. E. I. R.
Elmsdale		Prince	
Station	124		P. E. I. R.
Summit, 2·5 miles, E.	155		P. E. I. R.
Emerald	123	Prince	P. E. I. R.
Five Houses	125	Kings	P. E. I. R.
Fredericton		Queens	
Station	306		P. E. I. R.
Summit, 0·1 mile, W.	307	*la plus haut de même*	P. E. I. R.
Freetown		Prince	
Station	102		P. E. I. R.
Depression, 1·7 mile, W.	55		P. E. I. R.
Georgetown		Kings	
Station	26		P. E. I. R.
Wharf	6		P. E. I. R.
Harmony	161	Kings	P. E. I. R.
Harper	67	Prince	P. E. I. R.
Hazelbrook		Queens	P. E. I. R.
Howlan (Mill River)		Prince	
Station	55		P. E. I. R.
Summit, 0·5 mile, E.	111		P. E. I. R.
Hunter River	86	Queens	P. E. I. R.
Kensington	119	Prince	P. E. I. R.
Kildare (now St. Louis)		Prince	
Station	142		P. E. I. R.
Summit, 0·5 mile, W.	168		P. E. I. R.
Kinkora	111	Prince	P. E. I. R.
Lot 40	51	Kings	P. E. I. R.
Loyalist Road	48	Queens	P. E. I. R.
Marie	23	Kings	P. E. I. R.
Milton	25	Queens	P. E. I. R.

DEPARTMENT OF THE INTERIOR

PRINCE EDWARD ISLAND

Locality	Elev.	County	Authority
Miscouche	85	Prince	P. E. I. R.
Morelle	39	Kings	P. E. I. R.
Mount Stewart	24	Queens	P. E. I. R.
New Annan	50	Prince	P. E. I. R.
New Zealand	183	Kings	P. E. I. R.
Northam	74	Prince	P. E. I. R.
O'Leary	99	Prince	P. E. I. R.
Peake	110	Kings	P. E. I. R.
Perth	136	Kings	P. E. I. R.
Pisquid	59	Queens	P. E. I. R.
Point	159	Queens	P. E. I. R.
Portage	37	Prince	P. E. I. R.
Port Hill station	49	Prince	P. E. I. R.
Richmond		Prince	
Station	75		P. E. I. R.
Summit, 1·1 mile, W	132		P. E. I. R.
Rollo Bay	134	Kings	P. E. I. R.
Royalty junction	165	Queens	P. E. I. R.
St. Andrews	15	Kings	P. E. I. R.
St. Eleanor	80	Prince	P. E. I. R.
St. Louis	142	Prince	P. E. I. R.
St. Nicholas	53	Prince	P. E. I. R.
St. Peter	15	Kings	P. E. I. R.
Selkirk	101	Kings	P. E. I. R.
Souris		Kings	
Station	15		P. E. I. R.
Wharf	7		P. E. I. R.
Suffolk		Queens	
Station	156		P. E. I. R.
Summit, 0·9 mile, E.	173		P. E. I. R.
Summerside	8	Prince	P. E. I. R.
Tanks	24	Prince	P. E. I. R.
Tignish	62	Prince	P. E. I. R.
Tracadie	61	Queens	P. E. I. R.
Travellers Rest	36	Prince	P. E. I. R.
Union Road	118	Queens	P. E. I. R.
Village Green	116	Queens	P. E. I. R.
Wellington	10	Prince	Geol Surv.
Wiltshire (summit)	305	Queens	P. E. I. R.
Winsloe Road	144	Queens	P. E. I. R.
York	138	Queens	P. E. I. R.

QUEBEC

Locality	Elev.	County	Authority
Abbotsford	209	Rouville	C. P. R.
Abercorn		Brome	
Station	48,		C. P. R.
International boundary, 1·4 mile, S.	491		C. P. R.
Abitibi lake	830	Abitibi	O'Sullivan
Actonvale.		Bagot	
Grand Trunk station	311		G. T. R.
Can. Pac. station	317		C. P. R.
Can. Pac. Ry. crossing	318		G. T. R.
Adamsville	376	Brome	C. P. R.
Adirondack junction	99	Laprairie	C. P. R.
Agotawekami lake	875	Abitibi	O'Sullivan
Albert, Mount	3,560	Gaspé	Geol. Surv.
Allen Mills.		Portneuf	
Station	564		Q. & L. St. J. R.
Jacquot river, 0·8 mile, N., water, 551 ; rail	561		Q. & L. St. J. R.
Allumette or Pembroke lake, high water, 370 ; low water	363	Pontiac	Rys. & Canals
Amqui.	532	Matane	I. C. R.
Ange Gardien	17	Montmorency	Q. R. L. & P. Co.
Anse à Giles, hill five miles S.S.E. of	1,140	L'Islet	Admir. chart
Anse aux Vaches, peak one and half miles W. of	1,890	Montmagny	Admir. chart
Anticosti island.		Saguenay	
Summit of island	700		Admir. chart
Bear head	400		Admir. chart
Table head	290		Admir. chart
Macastey ridge	400		Admir. chart
High cliff	500		Admir. chart
West cliff	400		Admir. chart
Argenteuil	113	St. Hyacinthe	Q. S. R.
Arthabaska.		Arthabaska	
Station	429		G. T. R.
East branch Nicolet river, 0·7 mile, W., water, 389 ; rail	419		G. T. R.
Ascot.		Sherbrooke	
Station	631		Q. C. R.
Moulton Hill summit, 2·7 miles, W.	710		Q. C. R.
Ashpabanka lake	1,118	Abitibi	O'Sullivan
Ashwapmuchuan lake	1,100	L. St. John	O'Sullivan
Asinitchibastat lake	1,075	Abitibi	O'Sullivan
Askitichi lake	1,189	L. St. John	O'Sullivan
Assametquaghan	282	Bonaventure	I. C. R.
Assiwanan lake	1,220	Champlain	O'Sullivan
Aston	290	Nicolet	I. C. R.
Athelstan.		Huntingdon	
Station	172		N. Y. C. R.
International boundary, 4·9 miles, S.	266		N. Y. C. R.
Attikonak lake	1,700	Ashuanipi	Geol. Surv.
Aubrey	141	Châteauguay	C. A. R.
Auckland	1,532	Compton	Me. C. R.
Ayers Flat	554	Stanstead	B. & M. R.
Aylmer	217	Wright	P. P. J. R.
Aylmer, Lake, high water	816	Wolfe	Q. C. R.
Aylwin	493	Wright	O. N. & W. R.
Bagot	241	Bagot	I. C. R.
Baie du Febvre, B. M., in western tower of R. C. church.	77	Yamaska	Public Works
Bare mountain	1,520	Sherbrooke	Geol. Surv.
Barn mountain	3,400	Gaspé	Geol. Surv.
Barrington		Huntingdon	
Station	176		G. T. R.
Can. Atl. Ry. crossing	190		G. T. R.
Bastien lake	877	Pontiac	Geol. Surv.
Batiscan.		Champlain	
Station	37		C P. R.
Batiscan river, water, 35 ; rail	42		C. P. R.
Brunelle's wharf, floor level	17·8		Public Works
Champlain river, at road bridge, flood (1865)	28·0		Public Works
River St. Lawrence:—			
Standard low water	5·8		Public Works
Standard high water	23·6		Public Works

QUEBEC

Locality	Elev.	County	Authority
Batiscan..		Champlain	
River St Lawrence—			
Flood (1865).............................	28·0		Public Works
Flood (1885).............................	25·1		Public Works
Battle lake...................................	754	Wright	Geol. Surv.
Bayfield, Mount...........................	3,471	Matane	Geol. Surv.
Bayview.......................................		Jacques Cartier	
Grand Trunk station...................	114		G. T. R.
Can. Pac. station........................	115		C. P. R.
Beaconsfield................................		Jacques Cartier	
Grand Trunk station...................	105		G. T. R.
Can. Pac. station........................	108		C. P. R.
Beauce...		Beauce	
Station...................................	491		Q. C. R.
Chaudiere river, high water, 486; rail..	491		Q. C. R.
Beauce tank.................................	487	Beauce	Q. C. R.
Beauchêne...................................	548	Pontiac	C. P. R.
Beaudet.......................................		Portneuf	
Station...................................	882		Q. & L. St. J. R.
Batiscan river, 1·4 mile, N., water, 959; rail..	981		Q. & L. St. J. R.
Beauharnois................................	132	Beauharnois	N. Y. C. R.
Beauport.....................................	22	Quebec	Q. R. L & P. Co.
Beauport hills..............................	1,950	Quebec	Admir. chart
Beauport Road............................	26	Quebec	Q. & L. St. J. R.
Beaurepaire.................................		Jacques Cartier	
Grand Trunk station...................	105		G. T. R.
Can. Pac. station........................	106		C. P. R.
Beau Rivage (Pleasant Beach).........	366	Bonaventure	I. C. R.
Bécancour...................................		Nicolet	
River St. Lawrence, extreme high water (1865)...	29·0		Public Works
B.M., in top course of foundation wall of R.C. church	37·4		Public Works
Bedford.......................................	177	Missisquoi	C. P. R.
Bélair...	91	Jacques Cartier	C. P. R.
Bélair...	189	Portneuf	C. P. R.
Belisle Mill..................................	1,059	Terrebonne	C. P. R.
Bellay, Lac à...............................	815	L. St. John	Q. & L. St. J. R.
Bellevue......................................	98	Châteanguay	N. Y. C. R.
Beloeil..		Verchères	
Station...................................	62		G. T. R.
B. M., in west abutment of railway bridge......	47·2		Public Works
Berry mountains, Big....................	2,000	Matane	Geol. Surv.
Berry mountains, Little.................	1,500	Matane	Geol. Surv.
Berthier junction..........................	40	Berthier	C. P. R.
Berthier station............................	31	Berthier	C. P. R.
Bic highlands...............................	1,236	Rimouski	Admir. chart
Bic...		Rimouski	
Station...................................	82		I. C. R.
Little Bic river, water, 38; rail........	82		I. C. R.
Birch lake...................................	862	Pontiac	Geol. Surv.
Birchton.....................................	971	Compton	C. P. R.
Black Cape..................................		Bonaventure	
Station...................................	80		A. & L. S. R.
Summit, 1·5 mile, E., ground, 144; rail.........	134		A. & L. S. R.
Black Lake station........................	940	Megantic	Q. C. R.
Black River siding........................	899	Quebec	Q. & L. St. J. R.
Blake siding................................	268	Drummond	I. C. R.
Blue Sea lake, high water, 537; low water...........	533	Wright	O. N. & W. R.
Bolton Centre..............................		Brome	
Old station..............................	768		M. & B. R. V. R.
North branch Missisquoi river, water, 762; rail....	776		M. & B. R. V. R.
Bolton Mines, old station.............	812	Brome	M. & B. R. V. R.
Bonaventure................................	64	Bonaventure	A. & L. S. R.
Bonaventure island.......................	400	Gaspé	Admir. chart
Boucherville................................		Chambly	
Station...................................	41		S. S. R.
B. M. in S. W. corner of R. C. church......	48·4		Public Works
R. & O. Co.'s wharf, floor..............	28·8		Public Works
River St. Lawrence—			
Standard low water..................	17·3		Public Works
Extreme high water.................	38·2		Public Works

QUEBEC.

Locality	Elev.	County	Authority
Boulogne	241	Drummond	C. P. R.
Boundary	1,852	L. Megantic	C. P. R.
Bourg Louis		Portneuf	
Station	516		Q. & L. St. J. R.
Portneuf river, bed, 489 ; rail	513		Q. & L. St. J. R.
Bradore hills	1,264	Saguenay	Admir. chart
Branch lake	1,162	L. St. John	O'Sullivan
Bras-coupé lake	983	Abitibi	O'Sullivan
Breckenridge	215	Wright	P. P. J. R.
Brennan	365	Wright	O. N. & W. R.
Brigham station	297	Brome	C. P. R.
Brigham junction		Brome	
Station	289		C. P. R.
South branch Yamaska river, water, 243 ; rail	266		C. P. R.
Bristol station	477	Ponti.	P. P. J. R.
Bristol mines	314	Ponti.	P. P. J. R.
Bristol Mines junction	383	Pontiac	P. P. J. R.
Britannia Mills	221	Bagot	G. T. R.
Broadlands	36	Bonaventure	A. & L. S. R.
Brome Corner		Brome	
Station	678		C. P. R.
South branch Yamaska river, 1·1 mile S., water (July, 1892), 648 ; rail	661		C. P. R.
Brome lake (July 7, 1892)	649	Brome	C. P. R.
Brompton		Richmond	
Station	470		G. T. R.
St. Francis river, 1·4 mile, N., water, 416 ; rail	442		G. T. R.
Brompton lake	770	Sherbrooke	Geol. Surv.
Brookbury	716	Compton	Me. C. R.
Brosseau	56	Laprairie	G. T. R.
Brosseau Mill	422	Lothiniere	L. & M. R.
Broughton station	1,106	Beauce	Q. C. R.
Broughton tank	1,141	Beauce	Q. C. R.
Bryson		Châteauguay	
Station	135		G. T. R.
Châteauguay river, high water, 117 ; low water, 110 ; rail	135		G. T. R.
Buckingham station	183	Labelle	C. P. R.
Buckingham village, road crossing	437	Labelle	C. P. R.
Buckingham landing	423	Labelle	C. P. R.
Bulwer	930	Compton	C. P. R.
Bury	1,065	Compton	C. P. R.
Cabane Ronde	61	L'Assomption	C. P. R.
Cacouna	259	Temiscouata	I. C. R.
Calumet		Argenteuil	
Station	194		C. P. R.
Calumet river, bed, 141 ; rail	179		C. P. R.
Rouge river, 2·2 miles, W., high water (May 1876), 140 ; ordinary low water, 126 ; rail	169		C. P. R.
Camille mountain	2,036	Rimouski	Admir. chart
Campbell Bay		Pontiac	
Station	361		P. P. J. R.
Stevenson creek, high water (1878), 351 ; ordinary high water, 349 ; low water, 339 ; rail	354		P. P. J. R.
Canoe lake	1,405	L. St. John	O'Sullivan
Cap de la Magdeleine mills	49	Champlain	C. P. R.
Cape de la Baie, peak one mile N.W	2,067	Charlevoix	Admir. chart
Cape Dogs, peak behind	1,395	Charlevoix	Admir. chart
Capelton	493	Sherbrooke	B. & M. R.
Caplan	92	Bonaventure	A. & L. S. R.
Caplan River station	54	Bonaventure	A. & L. S. R.
Cap St. Ignace station	132	Montmagny	I. C. R.
Carbuncle mountain	1,270	Sherbrooke	Geol. Surv.
Caribou lake	1,292	Quebec	Q. & L. St. J. R.
Carleton	57	Bonaventure	A. & L. S. R.
Carleton, Mount	1,830	Bonaventure	Admir. chart
Carmel (summit)	349	Drummond	I. C. R.
Caroline	121	St. Hyacinthe	Q. S. R.
Carr	185	Huntingdon	G. T. R.

QUEBEC

Locality	Elev.	County	Authority
Cascades		Wright	
Station	304		O. N. & W. R.
Summit, 1·2 mile, S....	303		O. N. & W. R.
Cascapedia	48	Bonaventure	A. & L. S. R.
Caughnawaga....		Châteauguay	
B. M., in abutment of C. P. Ry. bridge............	72·8		Public Works
Chisel mark on coping of abutment of C. P. Ry. bridge	129·5		Public Works
Causapscal ...		Matane	
Station....	454		I. C. R.
Matapedia river, water, 425 ; rail............	454		I. C. R.
Cavaignac....	118	Bagot	C. P. R.
Cecile........	150	Beauharnois	C. A. R.
Cedar Hall.......	541	Matane	I. C. R.
Cedar lake.... ...... •	1,280	Quebec	Q. & L. St. J. R.
Cedars , ...		Soulanges	
Station...	161		G. T. R.
Top step, east side of R. C. church...	157·1		D. W. Com.
Chambly..		Chambly	
B. M., in S.W. corner of lockmaster's office.......	47·8		Public Works
B. M., in south abutment of C.V. R. R. bridge....	79·6		Public Works
B. M., in lower face of wall of lock 7............	47·9		Public Works
Top of cap of bench well, on south side of lock 7..	47·6		Public Works
Lower sill of lock 9	11·3		Public Works
Upper sill of lock 2........................	85·4		Public Works
Chambly basin....		Chambly	
Lowest normal level.......................	18·9		Public Works
Extreme low water.......................	18·3		Public Works
Extreme high water, (1869).................	35·0		Public Works
Chambord junction	550	L. St. John	Q. & L. St. J. R.
Champlain....		Champlain	
Station.....................	51		C. P. R.
Champlain river, bed, 17 ; rail.....	45		C. P. R.
Gagnon's wharf, floor level................	18·5		Public Works
River St. Lawrence—			
Standard low water...................	6·9		Public Works
Standard high water.	24·3		Public Works
Extreme high water.	28·1		Public Works
Champlain, Lake..		Missisquoi	
Standard high water (May 4, 1869), highest on record	101·0		Public Works
Standard low water................	92·8		Public Works
Extreme low water (Oct. 12, 1880).	92·1		Public Works
Mean water (1871 to 1895).................	94·9		Public Works
Charette Mill		St. Maurice	
Station....	403		G. N. R.
West branch Yamachiche river, 2·1 miles, E., high			
water, 355 ; low water, 346 ; rail.....	388		G. N. R.
River du Loup, 2·1 miles, W,, water, 357 ; rail....	484		G. N. R.
Charlesbourg	118	Quebec	Q. & L. St. J. R.
Charlesbourg West....		Quebec	
Station.............................	282		Q. & L. St. J. R.
River des Mères, 0·9 mile, W., bed, 327 ; rail......	347		Q. & L. St. J. R.
Châteauguay....		Châteauguay	
Station.................................	85		N. Y. C. R.
Châteauguay river, low water, 68; high water, 78; rail	86		N. Y. C. R.
Chateau Richer.................	17	Montmorency	Q. R. L. & P. Co.
Chateau Richer hills..........	2,305	Montmorency	Admir. chart
Chats lake, high water, 246 ; low water.	239	Pontiac	Rys. & Canals
Chaudière Curve............	231	Lévis	G. T. R.
Chaudière....		Lévis	
Station....	235		G. T. R.
Chaudière river, water (Sept. 4, 1878) 176 ; rail. ...	234		G. T. R.
Chaudière junction..........	234	Lévis	G. T. R.
Chegobish lake	1,106	L. St. John	O'Sullivan
Chelsea		Wright	
Station.....................	365		O. N. & W. R.
Summit, 0·5 mile, N..	395		O. N. & W. R.
Depression, 2·2 miles, N., rail....	288		O. N. & W. R.
Chevreuil, Lac..........................	1,053	Argenteuil	M. & G. C. R.
Chibougamau lake	1,152	Abitibi	O'Sullivan

QUEBEC

Locality	Elev.	County	Authority
Chicoutimi...		Chicoutimi	
Station...	19		Q. & L. St. J. R.
Chicoutimi river, 1·5 mile, W., water, 34 ; rail.....	89		Q. & L. St. J. R.
Chlorydorme mountain...	1,000	Gaspé	Admir. chart
Christopherson lake...	950	Abitibi	O'Sullivan
Clair, Lake...	1,246	Mistassini	O'Sullivan
Clarenceville...	122	Missisquoi	Q. S. R.
Clark...	588	Pontiac	P. P. J. R.
Clearwater lake...	1,326	L. St. John	O'Sullivan
Clifton...		Compton	
Station...	1,124		Me. C. R.
Clifton river, 0·8 mile, S., bed, 937 ; rail...	951		Me. C. R.
Coaticook...	1,006	Stanstead	G. T. R.
Coleraine...	869	Megantic	Q. C. R.
Commissioner Lake...		L. St. John	
Station...	1,275		Q. & L. St. J. R.
Summit, 3·8 miles, S...	1,330		Q. & L. St. J. R.
Como...	99	Vaudreuil	C. P. R.
Compton...	733	Compton	G. T. R.
Conception...	722	Labelle	C. P. R.
Cone mountain...	1,370	Labelle	Geol. Surv.
Conical mountain...	1,910	Gaspé	Geol. Surv.
Contrecœur...		Verchères	
Station...	65		S. S. R.
B. M., in front of R. C. church...	43·5		Public Works
R. & O. Co.'s wharf, floor level...	26·5		Public Works
River St. Lawrence—			
Standard low water...	13·1		Public Works
Standard high water...	28·5		Public Works
Extreme high water...	35·9		Public Works
Convent...	115	Jacques Cartier	G. T. R.
Cookshire...		Compton	
Can. Pac. station...	684		C. P. R.
Maine Central station...	679		Me. C. R.
Eaton river, bed, 654 ; rail...	675		C. P. R.
Maine Central R. R. crossing...	648		C. P. R.
Coteau du Lac, B. M., in vestry of R. C. church...	156·1	Soulanges	D. W. Com.
Coteau Junction...		Soulanges	
Station...	160		G. T. R.
B. M., south-west corner of bridge-seat, east abutment of C. A. Ry. bridge over River Delisle...	158·9		D. W. Com.
B.M., coping, east abutment of G. T. Ry. bridge over River Delisle...	160·4		D. W. Com.
Coteau Landing, B. M., in S. abutment of C. A. Ry. bridge over Soulanges canal...	161·1	Soulanges	D. W. Com.
Côte de Liesse...	139	Jacques Cartier	G. T. R.
Côte Ste. Thérèse...	154	Napierville	G. T. R.
Coudres island, summit...	390	Charlevoix	Admir. chart
Coulonge...		Pontiac	
Station...	361		P. P. J. R.
Coulonge river, 1·8 mile, W., extreme high water, 353·5 ; ordinary high water, 349 ; low water (Oct. 1896), 338 ; rail...	366		P. P. J. R.
Cowansville...	377	Missisquoi	C. P. R.
Craig Road...	335	Lévis	G. T. R.
Cross Point...	30	Bonaventure	A. & L. S. R.
Dablon...	980	L. St. John	Q. & L. St. J. R.
Dalhousie Mills...	225	Soulanges	C. P. R.
Danby...	437	Drummond	G. T. R.
Danville...		Richmond	
Station...	446		G. T. R.
Southwest branch Nicolet river, 1·4 mile, E., water (Aug. 20, 1878), 361 ; rail...	391		G. T. R.
David lake...	869	Pontiac	Geol. Surv.
De Quen...	834	L. St. John	Q. & L. St. J. R.
Deschambault...		Portneuf	
Station...	182		C. P. R.
Wharf, floor level...	13·7		Public Works.

QUEBEC

Locality	.	County	Authority
Deschambault....		Portneuf	
St. Lawrence river, standard low water	−4·7		Public Works
St. Lawrence river, bed, 1884..........	21·6		Public Works
Deschênes	204	Wright	C. P. R.
Deschênes lake, high water, 198; low water......	189	Wright	Rys. & Canals
Desert............	561	Wright	O. N. & W. R.
Des Plantes	489	Beauce	Q. C. R.
Devil lake	834	Pontiac	Geol. Surv.
Dewittville	158	Huntingdon	G. T. R.
Dillonton junction..............	812	Brome	M. & B. R. V. R.
Dinahogama lake..........	1,180	Abitibi	O'Sullivan
D'Israeli............		Wolfe	
Station............	829		Q. C. R.
D'Israeli bridge, high water, 817; rail........	829		Q. C. R.
Ditchfield	1,476	L. Megantic	C. P. R.
Dixie	90	Jacques Cartier	G. T. R.
Dixville............		Stanstead	
Station............	1,126		G. T. R.
Boundary Line creek, 4·4 miles, S., water,1,209; rail	1,250		G. T. R.
International boundary, 4·4 miles, S.......	1,250		G. T. R.
Doda lake..........	974	Abitibi	O'Sullivan
Dodge lake..........	625	Labelle	Geol. Surv.
Dominion	89	Jacques Cartier	G. T. R.
Donaldson lake........	336	Labelle	Geol. Surv.
Doncaster.........	1,244	Compton	Me. C. R.
Doré lake.........	1,127	Abitibi	O'Sullivan
Dorval............		Jacques Cartier	
Can. Pac. station.........	88		C. P. R.
Grand Trunk station........	87		G. T. R.
Doucet Landing, outside wharf, floor level.......	18·5	Nicolet	Public Works
Douglas lake	852	Pontiac	Geol. Surv.
Douglastown mountain	1,500	Gaspé	Admir. chart
Drummondville....		Drummond	
Intercolonial station..........	290		I. C. R.
Can. Pac. station.........	206		C. P. R.
L'Avenir branch switch.........	206		C. P. R.
Intercolonial Ry. crossing	290		C. P. R.
St. Francis river, high water, 256; low water, 248; rail....	261		I. C. R.
Dudswell station.........	686	Wolfe	Q. C. R.
Dudswell junction............		Wolfe	
Station...........	697		Q. C. R.
St. Francis river, 3·1 miles, S., water, 643; rail...	665		Me. C. R.
Duncan...........		Drummond	
Station	290		I. C. R.
Summit, 0·8 mile, E............	326		I. C. R.
Durnford.........	1,281	Terrebonne	C. P. R.
Eagle cape........	807	Charlevoix	Admir. chart
Eagle lake....	534	Rimouski	I. C. R.
Eardley	215	Wright	P. P. J. R.
East Angus........	674	Compton	Q. C. R.
East Broughton........		Beauce	
Station............	1,224		Q. G. R.
Summit, 0·7 mile, W.........	1,251		Q. C. R.
East Farnham.......		Brome	
Station	346		C. P. R.
South branch Yamaska river, 1·2 mile S., water, 304; rail....	329		C. P. R.
East Hereford station........	1,117	Compton	Me. C. R.
East Hereford gravel pit siding	1,164	Compton	Me. C. R.
Eastman........		Brome	
Can. Pac. station........	911		C. P. R.
Orford Mt. station..........	822		O. M. R.
Orford Mountain Ry. crossing, 1·6 mile E., O. M. Ry., rail, 822; Can. Pac. Ry., rail......	868		C. P. R.
Orford Mountain switch	909		C. P. R.
Summit, 1·5 mile, W..........	934		C. P. R.
East Templeton....		Wright	
Station	159		C. P. R.
Upper Blanche river, water, 138; rail........	150		C. P. R.

QUEBEC

Locality	Elev.	County	Authority
Eaton Corner........................	800	Compton	Me. C. R.
Echo lake...........................	905	Rimouski	I. C. R.
Echo Vale, road crossing	1,455	L. Megantic	C. P. R.
Edward, Lake.......................	1,196	Quebec	Q. & L. St. J. R.
Elgin Road	162	L'Islet	I. C. R.
Escuminac	26	Bonaventure	A. & L. S. R.
Etchemin, Lake.....................	1,213	Dorchester	Bdy. Com.
Eternity, Cape.....................	1,500	Charlevoix	Admir. chart
Eustis	499	Stanstead	B. & M. R.
Evans, Lake........................	612	Abitibi	O'Sullivan
Everett siding	700	Wolfe	Q. C. R.
Fame Point mountain...............	1,000	Gaspé	Admir. chart
Farndon............................	251	Missisquoi	C. P. R.
Farnham...........................		Missisquoi	
Station	192		C. P. R.
St. Guillaume branch switch........	204		C. P. R.
Central Vermont R. R. crossing, 0.2 mile, W.....	185		C. P. R.
Yamaska river, 0.4 mile, N., water (May, 1885), 186; rail	213		C. P. R.
Farrellton		Wright	
Station	345		O. N. & W. R.
Bay of Gatineau river, 2.4 miles, S., water, 322; rail	341		O. N. & W. R.
File-axe lake	1,470	L. St. John	O'Sullivan
Fitch siding	59	Levis	G. T. R.
Flodden		Richmond	
Station	880		O. M. R.
Summit, 1.2 mile, S	922		O. M. R.
Folliot lake	1,069	Rimouski	Bdy. Com.
Forestdale		Nicolet	
Station	291		I. C. R.
Gentilly river, 0.4 mile, E., bed, 285; rail.........	291		I. C. R.
Forrest lake	862	Pontiac	Geol. Surv.
Fort Ingalls.......................	561	Temiscouata	Tem. R.
Foster.............................	697	Brome	C. P. R.
François Xavier, Lake	1,185	Argenteuil	M. C. R.
Frechette Mill.....................	760	Wolfe	Q. C. R.
Fulford	585	Brome	C. P. R.
Garneau		Champlain	
Station	448		G. N. R.
River St. Maurice, 3.0 miles, W., high water, 300; low water, 290; rail........	333		G. N. R.
Garthby..		Wolfe	
Station	825		Q. C. R.
Garthby bridge, high water, 816; rail..........	820		Q. C. R.
Gaspé, Cape........................	692	Gaspé	Admir. chart
Gatineau...........................		Wright	
Station	174		C. P. R.
Gatineau river, 1.7 mile, W., high water (1876), 150; extreme low water, 125; rail	181		C. P. R.
Gentilly, B. M. in S. E. corner of R. C. church....	63.3	Nicolet	Public Works
Girard.............................	153	St. Johns	G. T. R.
Giroux.............................	25	Bonaventure	A. & L. S. R.
Glen Sutton.......		Brome	
Station	526		C. P. R.
International boundary, 2.9 miles, W....	512		C. P. R.
Golf Links.........................	97	Jacques Carter	C. P. R.
Goose Lake		Bonaventure	
Station.............................	174		A. & L. S. R.
Summit, 1.2 mile, W., ground, 225; rail.........	212		A. & L. S. R.
Gosford, Mount.....................	3,658	Beauce	Bdy. Com.
Gould (summit).....................	1,302	Compton	C. P. R.
Gracefield..........................		Wright	
Station	507		O. N. & W. R.
Pickanock river, 1.7 mile, S., water, 486; rail..	509		O. N. & W. R.
Grande Ligne.......................	143	St. Johns	G. T. R.
Grandes Piles.......................	345	Champlain	C. P. R.
Grand point, peak two miles W. of.....	2,650	Charlevoix	Admir. chart
Grand lake.........................	504	Wright	Geol. Surv.
Grand lake Victoria....	960	Pontiac	O'Sullivan

QUEBEC

Locality	Elev.	County	Authority
Great Beaver lake (Hair-cutting river)	1,415	Champlain	Geol. Surv.
Great Beaver lake (Rouge river)	843	Labelle	Geol. Surv.
Green lake (Rouge river)	864	Labelle	Geol. Surv.
Green lake (Blanche creek)	509	Labelle	Geol. Surv.
Grenville		Argenteuil	
Can. Pac. station	210		C. P. R.
Great Nor. station	176		G. N. R.
Grondines		Portneuf	
Station	128		C. P. R.
Floor level of w' arf	13·8		Public Works
R. C. church, top. of stone step at N. W. corner	41·9		Public Works
St. Lawrence river—			
Standard low water	−0·3		Public Works
Standard high water	+19·9		Public Works
Highest known flood	26·6		Public Works
Gros Visons, Lac	1,280	L. St. John	Q. & L. S. J. R.
Gubanne, Cape, peak 1½ miles N. W. of	2,215	Montmagny	Admir. chart
Gull lake	660	Abitibi	O'Sullivan
Hadlow	18	Lévis	G. T. R.
Hair-cutting lake	1,510	Champlain	Geol. Surv.
Hall siding	449	Megantic	G. T. R.
Ham mountain	1,950	Wolfe	Geol. Surv.
Hare island	323	Temiscouata	Admir. chart
Harlaka	240	Lévis	I. C. R.
Hébertville		L. St. John	
Station	515		Q. & L. St. J. R.
River Bédard, water, 492; rail	522		Q. & L. St. J. R.
Hedleyville	17	Quebec	Q. & L. St. J. R.
Height of land, St. Lawrence and Atlantic (Attikonak and Romaine rivers)	1,750	Saguenay	Geol. Surv.
Height of land, St. Lawrence & Hudson Bay			
Grand lake Victoria (Ottawa) and Nottaway river	1,000	Pontiac	Geol. Surv.
Du Chef and Mistassini waters	1,500	L. St. John	O'Sullivan
Nikabon and Obatogaman rivers	1,275	L. St. John	O'Sullivan
Hemmingford	224	Huntingdon	G. T. R.
Henryburg	205	Missisquoi	C. A. R.
Hentyville	116	Iberville	Q. S. R.
Hereford		Compton	
Station	1,085		Me. C. R.
International boundary crossing	1,089		Me. C. R.
Hillcrest	551	Pontiac	C. P. R.
Hillhurst	818	Compton	G. T. R.
Highlands	118	Jacques Cartier	C. P. R.
Hochelaga	63	Hochelaga	C. P. R.
Hogsback hill	2,381	Beauce	Bdy. Com.
Holton		Châteauguay	
Station (closed)	190		C. A. R.
B. M., in stone support of water tank 1½ mile E.	180·1		D. W. Com.
Howick station	130	Châteauguay	G. T. R.
Howick junction		Châteauguay	
Station	133		C. A. R.
B. M., west abutment of C. A. Ry. bridge over English river	126·3		D. W. Com.
B. M., west abutment of C. A. Ry. bridge over Chateauguay river	129·6		D. W. Com.
Hudson	92	Vaudreuil	C. P. R.
Hudson Heights	87	Vaudreuil	C. P. R.
Hull		Wright	C. P. R.
Can. Pac. station	189		
P. P. J. station	163		P. P. J. R.
Huntingdon		Huntingdon	
Grand Trunk station	165		G. T. R.
New York Cent. station	167		N. Y. C. R.
Iberville		Iberville	
Can. Pac. station	111		C. P. R.
Que. Sou. station	103		Q. S. R.
B. M. in S. W. corner of D. Frechette's store, cor. Napier and St. Ann sts	103·7		Public Works
Iberville junction	115	Iberville	C. P. R.

QUEBEC

Locality	Elev.	County	Authority
Indian Lorette		Quebec	
Station	72		Q. & L. St. J. R.
River St. Charles, bed, 444 ; rail	430		Q. & L. St. J. R.
Irishtown	165	Bonaventure	A. & L. S. R.
Ironside	182	Wright	O. N. & W. R.
Ishimanikuagan lake..........	685	Saguenay	Geol. Surv.
Island lake..........................	667	Temiscouata	I. C. R.
Isle Cadieux.......................	94	Vaudreuil	C. P. R.
Isle Verte......		Temiscouata	
Station	117		I. C. R.
River Verte, 0·7 mile, W., high water, 92 ; low water, 85 ; rail	125		I. C. R.
Ivry....	1,213	Terrebonne	C. P. R.
Jacopie lake	1,630	Ashuanipi	Geol. Surv.
Jacques Cartier junction........	96	Jacques Cartier	G. T. R.
Johnson or Mistikus lake	1,027	Rimouski	Bdy. Com.
Johnson (Barrington).................		Huntingdon	
Station	192		C. A. R.
Summit 2·0 miles, E...... ...	232		C. A. R.
Chambly River, bed, 172 ; rail......	183		C. A. R.
Grand Trunk. Ry. crossing......	190		C. A. R.
B. M., in wall of culvert at second road crossing E. of station	210·0		D. W. Com.
Johnville..........................	852	Compton	C. P. R.
Joliette....		Joliette	
Great Northern station	204		G. N. R.
Can. Pac. station..........	196		C. P. R.
L'Assomption river, water, 185 ; rail......	204		G. N. R.
Joliette junction	73	Joliette	C. P. R.
Joliette junction	208	Joliette	G. N. R.
Jonquière		Chicoutimi	
Station	484		Q. & L. St. J. R.
River aux Sables, high water, 452 ; low water, 447 ; rail..................	480		Q. & L. St. J. R.
Jordan lake.........................	1,156	L. St. John	O'Sullivan
Juggler mountain	1,590	Abitibi	O'Sullivan
Kapadigoitch lake...................	1,192	L. St. John	O'Sullivan
Kapitachuan lake....	1,290	Pontiac	O'Sullivan
Kazabazua		Wright	
Station	571		O. N. & W. R.
Kazabazua creek, bed, 517 ; rail......	561		O. N. & W. R.
Summit 0·3 mile, N..................	599		O. N. & W. R.
Keen (closed)	1,727	L. Megantic	C. P. R.
Kelvin, Lake	415	Abitibi	O'Sullivan
Kempt.........		Matane	
Station	688		I. C. R.
Summit, 0·5 mile, S......	716		I. C. R.
Kendall lake........................	694	Labelle	Geol. Surv.
Kenogami...........................	546	Chicoutimi	Q. & L. St. J. R.
Kikendatch.........................	1,200	Champlain	O'Sullivan
King siding	53	Kamouraska	I. C. R.
Kings mountain.....................	1,085	Wright	Geol. Surv.
Kingsburg	384	Lotbinière	I. C. R.
Kingsbury	549	Richmond	O. M. R.
Kingsey	443	Richmond	G. T. R.
Kingsmere lake	725	Wright	Geol. Surv.
Kingsville.........................	1,026	Megantic	Q. C. R.
Kipawa station.....	885	Pontiac	C. P. R.
Kipawa junction	580	Pontiac	C. P. R.
Kipawa lake, high water, 883 ; low water....	873	Pontiac	C. P. R.
Kirks Ferry		Wright	
Station	294		O. N. & W. R.
Bay of Gatineau river, 0·4 mile, N., water, 286 ; rail.	296		O. N. & W. R.
Kiskisink.........................		Quebec	
Station	1,306		Q. & L. St. J. R.
Kiskisink river, high water, 1,301 ; low water, 1,299; rail	1,307		Q. & L. St. J. R.
Summit, 3·9 miles, N., between St. Maurice and Lake St. John waters....	1,431		Q. & L. St. J. R.

QUEBEC

Locality	Elev.	County	Authority
Knowlton.........		Brome	
Station	681		C. P. R.
North branch Yamaska river, water (July, 1892), 651; rail...	681		C. P. R.
Kowpettap lake..................	1,240	Champlain	Geol. Surv.
La Baie village	80	Yamaska	A. & L. S. R.
Labelle	749	Labelle	C. P. R.
Laberge Mill....	1,345	Terrebonne	C. P. R.
L'Acadie...		St. Johns	
Can. Pac. station........	131		C. P. R.
Grand Trunk station.....	118		G. T. R.
Little Montreal river, bed, 98; rail............	126		C. P. R.
Lachevrotière	107	Portneuf	C. P. R.
Lachine.....		Jacques Cartier	
Station	131		G. T. R.
B. M., in bridge over old waterworks canal..	72·1		Public Works
B. M., in Can. Pac. Ry. bridge over lower Lachine road...	93·4		Public Works
B. M., in S. E. corner of R. C. church...	82·3		Public Works
Lachute..............		Argenteuil	
Great Northern station.................	208		G. N. R.
Can. Pac. station.....	229		C. P. R.
North river, high water (1899), 205; rail.	213		G. N. R.
Lachute junction..............	250	Argenteuil	G. N. R.
Lacolle		St. Johns	
Grand Trunk station....	133		G. T. R.
Can. Atlantic station	152		C. A. R.
Richelieu river, high water, 101·4; low water, 92·7; rail.....	105		C. A. R.
B. M., in ballast wall, east abutment, Can. Atl. Ry. bridge over Lacolle river.....	165·3		D. W. Com.
Lacolle junction.	131	Missisquoi	C. A. R.
Lac Simon siding........	832	Portneuf	Q. & L. St. J. R.
Lac à Tortue...		Champlain	
Station	439		C. P. R.
Summit, 0·7 mile, S....	449		C. P. R.
La Fourche	255	Montcalm	G. N. R.
Lady Beatrix lake	703	Abitibi	O'Sullivan
Lafontaine lake, high water, 299; low water..........	291	Pontiac	Rys. & Canals
Lake Bouchette station..........	1,136	L. St. John	Q. & L. St. J. R.
Lake Caribou		Portneuf	
Station	611		G. N. R.
Discharge of Caribou lake, water, 608; rail........	612		G. N. R.
Summit, 1·4 mile, E., ground, 737; rail	725		G. N. R.
Lake Edward...		Quebec	
Station..	1,195		Q. & L. St. J. R.
North-east branch Batiscan river, 2·3 miles, S., water, 1,186; rail. ..	1,189		Q. & L. St. J. R.
Lake Edward siding (summit).............	1,495	Quebec	Q. & L. St. J. R.
Lake aux Sables.............		Portneuf	
Station	514		G. N. R.
River Propre, 1·4 mile, W., b"d, 490; rail........	521		G. N. R.
Lakeside..........		Jacques Cartier	
Grand Trunk station	97		G. T. R.
Can. Pac. station...	95		C. P. R.
Lake St. Joseph station..............	534	Portneuf	Q. & L. St. J. R.
Lake Sergent.		Portneuf	
Station	550		Q. & L. St. J. R.
Summit, 2·8 miles, E....	632		Q. & L. St. J. R.
Lake siding..........	673	Sherbrooke	C. P. R.
Lake Park siding.	660	Sherbrooke	C. P. R.
Lake Weedon station..............	830	Wolfe	Q. C. R.
Lambton.		Beauce	
Station	1,046		Q. C. R.
Summit, 2·3 miles, N.	1,128		Q. C. R.
L'Ange Gardien East..........		Rouville	
Station	235		C. P. R.
Summit, 0·8 mile, N	243		C. P. R.
L'Ange Gardien West.....	183	Labelle	C. P. R.

QUEBEC

Locality	Elev.	County	Authority
Lanoraie	76	Joliette	C. P. R.
Lanoraie	81	Richelieu	S. S. R.
L'Anse à Giles.....	118	L'Islet	I. C. R.
La Plaine.	188	Terrebonne	C. P. R.
Laprairie		Laprairie	
Station	57		G. T. R.
B. M., in N. W. corner of R. C. church......... ..	55·9		Public Works
Floor level of R. & O. Co.'s wharf.............. ..	36·7		Public Works
River St. Lawrence—			
Lowest known water during season of navigation	33·1		Public Works
Highest known water during season of navigation..............	37·8		Public Works
Laurentides........................	712	Portneuf	Q. & L. St. J. R.
Lavaltrie	89	Joliette	C. P. R.
Lavigne.	105	Vaudreuil	C. P. R.
Lawrenceville..............................		Shefford	
Station	705		O. M. R.
Black river, S.W. of station, water......	692		O. M. R.
Lennoxville..............................		Sherbrooke	
Grand Trunk station...............:..... ...	499		G. T. R.
Can. Pac. station........................	498		C. P. R.
Massawippi river, water (June, 1878), 473 ; rail. ...	497		G. T. R.
Can. Pac. Ry.; crossing.....	498		G. T. R.
L'Epiphanie........................	76	L'Assomption	C. P. R.
Le Rocher..	494	Beauce	Q. C. R.
Lesage...........................	586	Terrebonne	C. P. R.
Les Eboulements, church...............	1,186	Charlevoix	Admir. chart
Les Eboulements, Mount.....	2,551	Charlevoix	Admir. chart
L'Etoile................................		Portneuf	
Station	513		G. N. R.
Batiscan river, high water, 475 ; rail...............	487		G. N. R.
Lévis.............		Lévis	
Station	15		I. C. R.
Lévis hills......	371		Admir. chart
Graving dock, coping....................	17·1		Public Works
B. M., in rock, north end of I. C. Ha.. ·...	16·2		Public Works
B. M., in retaining wall, Geo. Couture's ...	22·3		Public Works
B. M., in rock, east side of Foisy's house	19·4		Public Works
River St. Lawrence—			
Extreme high water during season of navigation (Nov. 5, 1884).................	15·7		Public Works
Flood (Jan., 1838)	+45·4		Public Works
Standard low water.....	−8·4		Public Works
Standard high water......	+15·0		Public Works
Ordinary high water, high water season......	7·4		Public Works
Ordinary high water, low water season........	+6·2		Public Works
Ordinary low water, high water season.......	−3·1		Public Works
Ordinary low water, low water season..	−5·5		Public Works
Mean tide level, high water season	+2·2		Public Works
Mean tide level, low water season..	0·3		Public Works
Libby Mills	559	Stanstead	B. & M. R.
Lichen lake...........................	842	Abitibi	O'Sullivan
Lisgar........................	528	Drummond	G. T. R.
L'Islet................		L'Islet	
Station .... , ...	104		I. C. R.
Range of hills five miles S. E. of...............	1,390	L'Islet	Admir. chart
Little capes	1,000	Gaspé	Admir. chart
Little Metis	569	Matane	I. C. R.
Little River station..................	86	Vaudreuil	C. P. R.
Logan, Mount.........................	3,708	Matane	Geol. Surv.
Long lake (Broadback river)...............	673	Abitibi	O'Sullivan
Long lake..	1,367	Abitibi	Geol. Surv.
Long lake, low water..	820	Pontiac	C. P. R.
Long lake (Rouge river)..	854	Labelle	Geol. Surv.
Long pond.	812	Brome	M. & B. R. V. R.
Longueuil..............................		Chambly	
Station	55		S. S. R.
B. M. in N. E. side of R. C. church..............	50·1		Public Works
Floor level of new Govt. wharf	28·2		Public Works

QUEBEC

Locality	Elev.	County	Authority
Longueuil...............................		Chambly	
River St. Lawrence—			
Standard low water...............	18·5		Public Works
Standard high water..............	33·9		Public Works
Highest known flood.............	44·6		Public Works
Loon lake..............................	1,450	L. St. John	O'Sullivan
Loon lake..............................	1,345	Champlain	Geol. Surv.
Lorette................................		Quebec	
Station...........................	87		C. P. R.
Old Lake St. John junction, 2·6 miles, E........	50		C. P. R.
Lotbinière, B. M. in S. W. corner of R. C. church.....	88·9	Lotbinière	Public Works
Louiseville............................		Maskinongé	
Station...........................	46		C. P. R.
River du Loup, water, 24; rail........	48		C. P. R.
Louis, Mount..........................	1,000	Gaspé	Admir. chart
Lourdes...............................	412	Megantic	L. & M. R.
Low...................................		Wright	
Station...........................	415		O N. & W. R.
Summit, 2·4 miles, N., ground, 545; rail........	530		O. N. & W. R.
Lower Laurentian junction.............	494	Champlain	G. N. R
Lumsden Mill..........................	801	Pontiac	C. P. R.
Lyster................................		Megantic	
Station...........................	446		G. T. R.
Becancour river, water, 425; rail........	446		G. T. R.
Summit, 2·3 miles, N.............	472		L. & M. R.
Village...........................	463		L. & M. R.
Maddington Falls.......................		Nicolet	
Station...........................	281		I. C. R.
River Bécancour, bed, 242; rail........	281		I. C. R.
Magdalen Islands.......................		Gaspé	
Great Bird rock...................	105		Admir. chart
Bryon island......................	200		Admir. chart
Alright, Cape.....................	400		Admir. chart
Alright island....................	420		Admir. chart
Grindstone island.................	550		Admir. chart
Amherst island....................	550		Admir. chart
Demoiselle hill...................	280		Admir. chart
Entry island......................	580		Admir. chart
Grosse isle.......................	300		Admir. chart
Magog................................		Stanstead	
Station...........................	689		C. P. R.
Magog river, water (July 3, 1893), 682; rail....	692		C. P. R.
Magog lake, Little.....................	634	Sherbrooke	C. P. R.
Makustigan lake.......................	1,140	Abitibi	O'Sullivan
Mansonville...........................		Brome	
Can. Pac. station.................	516		C. P. R.
Road crossing at north end of village....	622		M. & B. R. V. R.
Mansonville pond, water...........	582		M. & B. R. V. R.
Marbleton.............................		Wolfe	
Station...........................	684		Q. C. R.
Marbleton bridge, bed, 661; rail........	671		Q. C. R.
Maria.................................	35	Bonaventure	A. & L. S. R.
Maria Capes station...................	127	Bonaventure	A & L. S. R.
Maria mountain........................	1,230	Bonaventure	Admir. chart
Marks Crossing........................	595	Wright	O. N. & W. R.
Mascouche............................		Terrebonne	
Station...........................	181		C. P. R.
Mascouche river, bed, 145; rail........	192		C. P. R.
Maskinongé lake.......................	694	Wright	Geol. Surv.
Maskinongé............................		Maskinongé	
Station...........................	53		C. P. R.
Maskinongé river, high water, 20; low water, 23; rail.	52		C. P. R.
Massawippi............................	536	Stanstead	B. & M. R.
Massawippi mountain...................	1,350	Brome	Geol. Surv.
Matapedia.............................		Bonaventure	
Station...........................	53		I. C. R.
Restigouche river, freshet water, 31; summer water, 13; rail.	45		I. C. R.
Matapedia, Lake.......................	515	Matane	I. C. R.

QUEBEC

Locality	Elev.	County	Authority
Matawee, Mount	3,365	Matane	Geol. Surv.
Matonipi lake	1,640	Saguenay	Geol. Surv.
Mattagami lake	615	Abitibi	O'Sullivan
Mattawagosik hills	1,590	Pontiac	Geol. Surv.
Mattawagosik lake	900	Abitibi	Geol. Surv.
McArthur lake	530	Labelle	Geol. Surv.
McConnell or Thompson lake	874	Pontiac	Geol. Surv.
McFee lake	661	Labelle	Geol. Surv.
McGregor lake	458	Wright	Geol. Surv.
McKee	521	Pontiac	P. P. J. R.
McKenzie siding	63	Temiscouata	I. C. R.
McLeod	1,469	Compton	C. P. R.
Meach lake	580	Wright	Geol. Surv.
Megantic		L. Megantic	
Station	1,314		C. P. R.
Chaudière river, bed, 1,295 ; rail	1,313		C. P. R.
Megantic lake, extreme high water	1,303	L. Megantic	Q. C. R.
Meguic		Portneuf	
Station	837		Q. & L. St. J. R.
Meguic river, water, 818 ; rail	832		Q. & L. St. J. R.
Mekattina highland	685	Saguenay	Admir. chart
Melbourne Ridge station	822	Richmond	O. M. R.
Mellon	365	Pontiac	P. P. J. R.
Melvina	1,442	Compton	Me. C. R.
Memphremagog, Lake	682	Stanstead	C. P. R.
Menihek lakes	1,750	Ashuanipi	Geol. Surv.
Metabetchouan	373	L. St. John	Q. & L. St. J. R.
Méthots Mills		Lotbiniere	
Station	444		G. T. R.
River du Chêne, water (Aug. 30, 1878), 420 ; rail	432		G. T. R.
Mikwasash lake	1,006	Abitibi	O'Sullivan
Milan (summit)	1,646	L. Megant.	C. P. R.
Nile End	225	Hochelaga	C. P. R.
Milletta	893	Stanstead	C. P. R.
Millstream		Bonaventure	
Station	139		I. C. R.
Matapedia river, 1·1 mile, N., water, 129 ; rail	158		I. C. R.
Mirror lake	1,423	Quebec	Q. & L. St. J. R.
Mistassini, Lake	1,200	Abitibi	O'Sullivan
Mistassinis, Lake	1,246	Mistassini	O'Sullivan
Mistikus lake	1,027	Rimouski	Bdy. Com.
Mitchell		Nicolet	
Station	227		T. C. R.
South-west branch Nicolet river, water, 191 ; bed, 187 ; rail	221		I. C. R.
Mohr		Pontiac	
Station	226		P. P. J. R.
Mohr creek, water, 214 ; rail	236		P. P. J. R
Montcalm		Montcalm	
Station	231		G. N. R.
Ouareau river, high water, 194 ; low water, 189 ; rail	224		G. N. R.
Montebello		Labelle	
Station	166		C. P. R.
Salmon river, 1·8 mile, E., high water (1876), 140 ; low water, 126 ; rail	156		C. P. R.
Montfort	1,204	Argenteuil	M. & G. C. R.
Montfort junction	528	Terrebonne	C. P. R.
Montmagny		Montmagny	
Station	55		I. C. R.
River du Sud, water, 49 ; rail	58		I. C. R.
Montmorency	18	Quebec	Q. R. L. & P. Co.
Montreal		Hochelaga	
Bonaventure station	47		G. T. R.
Windsor station	108		C. P. R.
Place Viger station, base of rail	57		C. P. R.
City Hall, floor level	103·7		City Eng.
City Hall, roof of main tower	237·4		City Eng.
Reservoir, McTavish street, bed	206		City Eng.
Mount Royal, summit	763		City Eng.

QUEBEC

Locality	Elev.	County	Authority
Montreal...		Hochelaga	
Victoria Jubilee bridge, at Montreal end, rail....	66·3		Public Works
City and Ordnance datum....................	23·9		City Eng.
Grand Trunk Ry. datum.........................	8·3		G. T. R.
Bench Marks—			
Ordnance B. M., front of Molson's distillery, Notre Dame street.............................	60·0		City Eng.
Foundation stone of house, N. W. corner St. Denis and Sherbrooke streets...............	137·5		City Eng.
Top of stone plinth, S. W. cor. Mount Royal and St. Lawrence streets...................	212·1		City Eng.
Stone post of wall, W. side of Côte des Neiges hill opposite Mr. D. R. McCord's.........	359·2		City Eng.
Rock, lower end and S. side of St. Helen island, W. side of road round island............	77·3		Public Works
Third course, E. face of S. abutment of C. P. Ry. bridge over Brock street..............	40·7		Public Works
Stone, E. corner of S. abutment of C. P. Ry. bridge over Forsyth street..................	37·2		Public Works
S. wall of custom-house, near cor. of Commissioner and Port streets....................	48·5		Public Works
First quoin above plinth angle of examining warehouse on Commissioner street........	48·4		Public Works
Lachine canal—			
Old lock No. 1, W. side of lower sill........	4·8		Public Works
Old lock No. 1, coping....................	36·2		Public Works
New lock No. 1, centre of lower sill........	2·9		Public Works
New lock No. 1, coping....................	35·8		Public Works
River St. Lawrence, at foot of lock No. 1—			
Extreme low water (Oct., 1895)...........	18·0		Public Works
Flood (1886)..............................	49·1		Public Works
Standard low water......................	20·8		Public Works
Standard high water.....................	35·5		Public Works
Montreal junction.............................	155	Hochelaga	C. P. R.
Montreal West................................	68	Jacques Cartier	G. T. R.
Moose hill....................................	2,862	Beauce	Bdy. Com.
Moose Park..................................		Nicolet	
Station..................................	301		I. C. R.
River du Chêne, water, 284; rail...........	301		I. C. R.
Summit, 0·6 mile, W......................	327		I. C. R.
Morrison.....................................	895	Terrebonne	C. P. R.
Moss-covered mountain.......................	1,650	Mistassini	O'Sullivan
Mount Tremblant station.....................	748	Terrebonne	C. P. R.
Murray mountains............................	2,300	Charlevoix	Admir. chart
Mushalagan lake..............................	830	Saguenay	Geol. Surv.
Mystic.......................................	180	Missisquoi	C. P. R.
Nantel.......................................	1,269	Terrebonne	C. P. R.
Narrow Ridge lake............................	1,164	L. St. John	O'Sullivan
Natoush lake.................................	1,340	Champlain	Geol. Surv.
Nemiskau lake................................	549	Mistassini	O'Sullivan
New Carlisle.................................	9	Bonaventure	A. & L. S. R.
New Erin.....................................	174	Huntingdon	N. Y. C. R.
New Glasgow.................................		Terrebonne	
Station..................................	240		G. N. R.
L'Assomption river, water, 231; rail......	245		G. N. R.
Newington...................................	527	Sherbrooke	Q. C. R.
New Liverpool................................	115	Lévis	G. T. R.
New Richmond...............................	53	Bonaventure	A. & L. S. R.
Newton lake.................................	666	Labelle	Geol. Surv.
Nicolet......................................		Nicolet	
Intercolonial station......................	75		I. C. R.
Quebec southern Ry. crossing.............	76		I. C. R.
Port St. Francis wharf, floor level.........	19·8		Public Works
B.M., in N.W. corner of R. C. cathedral......	68.3		Public Works
River St. Lawrence—			
Standard low water......................	9·5		Public Works
Standard high water.....................	26·5		Public Works
Extreme high water......................	31·4		Public Works

QUEBEC

Locality	Elev.	County	Authority
Nikabau lake	1,150	L. St. John	O'Sullivan
Normandin lake	1,275	L. St. John	O'Sullivan
North Derby	710	Stanstead	B. & M. R.
North Hatley	538	Stanstead	B. & M. R.
North Nation Mills		Labelle	
Station	181		C. P. R.
North Nation river, high water (1876), 141; rail	179		C. P. R.
North Stukely	734	Shefford	O. M. R.
North Wakefield	835	Wright	O. N. & W. R.
Notre Dame des Anges		Portneuf	
Station	551		G. N. R.
Summit, 2·2 miles W., ground, 636; rail	625		G. N. R.
Notre Dame du Lac	517	Temiscouata	Tem. R.
Nouvelle	53	Bonaventure	A. & L. S. R.
Nouvelle West		Bonaventure	
Station	80		A. & L. S. R.
Summit, 2·6 miles, W	238		A. & L. S. R.
Nouvelle mountain	1,058	Bonaventure	Admir. chart
Noyan	123	Missisquoi	C.A.R.
Oak Bay	64	Bonaventure	A. & L. S. R.
Obamiskachi lake	1,160	L. St. John	O'Sullivan
Obaska lake	882	Abitibi	O'Sullivan
Obashing lake	822	Pontiac	Geol. Surv.
Obashing lake, Little	832	Pontiac	Geol. Surv.
Obatagamau lake	1,120	Abitibi	O'Sullivan
Obikoba lake	852	Abitibi	Geol. Surv.
O'Donnell siding	984	L. St. John	Q. & L. St. J. R.
Old Lake Road	351	Temiscouata	I. C. R.
Old-man mountain	2,669	Gaspé	Geol. Surv.
Old-man lake	1,330	L. St. John	O'Sullivan
Olga lake	635	Abitibi	O'Sullivan
Opatawaga lake	702	Abitibi	O'Sullivan
Opawakau lake	836	Abitibi	O'Sullivan
Opemiska lake	1,031	Abitibi	O'Sullivan
Orford lake, high water, 923; low water	917	Brome	C. P. R.
Orford mountain	2,860	Sherbrooke	Geol. Surv.
Orleans island	500	Montmorency	Admir. chart
Ormstown	145	Châteauguay	G. T. R
Ossokmanuan Lake	1,600	Ashuanipi	Geol. Surv.
Otterburn	505	Temiscouata	Tem. R.
Otterburn Park	63	Rouville	G. T. R.
Ouiatchuan Falls station	346	L. St. John	Q. & L. St. J. R.
Ouiatchuan lake	1,017	L. St. John	Q. & L. St. J. R.
Outremont junction	195	Jacques Cartier	C. P. R.
Owen, Lake	1,174	L. St. John	O'Sullivan
Owls Head mountain	2,484	Brome	Royal Eng.
Paint mountain	1,400	Abitibi	O'Sullivan
Papineau	212	Rouville	C. P. R.
Papineauville	149	Labelle	C. P. R.
Paquetteville	1,291	Compton	Me. C. R.
Parisville	174	Lotbinière	L. & M. R.
Paskagama lake	1,112	Abitibi	O'Sullivan
Patamisk lake	1,800	Ashuanipi	Geol. Surv.
Patikamiki hill	1,250	Abitibi	Geol. Surv.
Patikamiki lake	744	Abitibi	O'Sullivan
Patrick mountain	1,630	L. St. John	O'Sullivan
Pearl Lake station	1,132	Quebec	Q. & L. St. J. R.
Pearl lake	1,125	Quebec	Q. & L. St. J. R.
Pemiskash lake	1,245	Champlain	O'Sullivan
Percée rock	288	Gaspé	Admir. chart
Percée or Table Roulante mountain	1,230	Gaspé	Admir. chart
Perthuis		Portneuf	
Station	651		Q. & L. St. J. R.
Rivière Noir, water, 631; rail	643		Q. & L. St. J. R.
Petitsikapau lake	1,675	Ashuanipi	Geol. Surv.
Piedmont	552	Terrebonne	C. P. R.
Pierreville village	76	Yamaska	A. & L. S. R.
Pike lake, low water	794	Pontiac	C. P. R.

QUEBEC

Locality	Elev.	County	Authority
Pikuzettizin lake	1,192	Abitibi	O'Sullivan
Piles junction	127	Champlain	C. P. P.
Pine lake	918	Argenteuil	M. & G. C. R.
Plessisville	441	Megantic	G. T. R.
Pointe à l'Abattis, peak two miles west of	2,530	Charlevoix	Admir. chart
Pointe au Chêne		Argenteuil	
Station	182		C. P. R.
River au Chêne, 0·6 mile. E. water, 157; rail	179		C. P. R.
Pointe Claire		Jacques Cartier	
Can. Pac. station	112		C. P. R.
Grand Trunk station	107		G. T. R.
Pointe du Lac	72	St. Maurice	C. P. R.
Pointe la Garde	49	Bonaventure	A. & L. S. R.
Pointe Lévis	16	Lévis	G. T. R.
Pointe Fortune	130	Vaudreuil	C. P. R.
Pointe Platon		Lotbinière	
St. Lawrence river, flood (1884)	19·6		Public Works
Wharf, floor level	14·4		Public Works
Point St. Charles	53	Hochelaga	G. T. R.
Pointe aux Trembles		Portneuf	
Grenier's wharf, floor level	13·6		Public Works
St. Lawrence river, flood (1884)	19·8		Public Works
Pont Rouge		Portneuf	
Station	356		C. P. R.
Jacques Cartier river, water, 288; rail	340		C. P. R.
Porcupine lake	1,337	Rimouski	I. C. R.
Porcupine mountain	2,125	Mistassini	O'Sullivan
Portage lake	1,590	Beauce	Bdy. Com.
Portneuf		Portneuf	
Station (summit)	197		C. P. R.
Portneuf river, water, 95; rail	169		C. P. R.
Presq'ile lake	1,059	Abitibi	O'Sullivan
Quaquakamaksis, Lake	1,307	L. St. John	Q. & L. St. J. R.
Quebec		Quebec	
Canadian Pacific station	18·8		C. P. R.
Quebec & Lake St. John station	18		Q. & L. St. J. R.
Ordnance datum	1·1		Public Works
Dufferin terrace and Chateau Frontenac	130		Baillargé
Citadel, flagstaff battery	360		Baillargé
Waterworks reservoir at Lorette	486		Baillargé
Perrault hill	330		Baillargé
West end of citadel at old French earthwork	330		Baillargé
St. Columba church (Point Pizeau)	144		Admir. chart
Spencer Grange	223		Admir. chart
Plains of Abraham	284		Admir. chart
Mount Pleasant	273		Admir. chart
Diamond Harbour martello tower	318		Admir. chart
St. John's church	305		Admir. chart
Zero of Harbour Commissioner's gauge, Pointe à Carcy wharf	--6·7		Public Works
Louise embankment, coping	17·3		Public Works
Bench Marks—			
Facade of Notre Dame de la Garde church	17·6		Public Works
E. corner of E. tower of gaol wall	288·9		Public Works
Ordnance bench mark, broad arrow on E. side of martello tower	287·4		Public Works
Ordonnance B. M. broad arrow on N. side of portico of Quebec bank	17·2		Public Works
Zero of barometer at observatory	292·1		Public Works
Quoin at N. E. angle front of Queen's store, Champlain st., about 3½ ft. above street	24·1		Public Works
Second course, stone plinth, E. facade of examining warehouse	16·6		Public Works
Quinze, Lac des	845	Pontiac	Geol. Surv.
Quio		Pontiac	
Station	275		P. P. J. R.
Quio river, high water, 242; low water, 239; rail	282		P. P. J. R.
Racine	897	Shefford	O. M. R.
Radnor	254	Champlain	C. P. R.

QUEBEC

Locality	Elev.	County	Authority
Red Spruce lake	1,073	Abitibi	O'Sullivan
Reids Camp		Portneuf	
Station	597		G. N. R.
River Tawachiche, 1·1 mile E., bed, 529; rail	554		G. N. R.
Rhéanme, Lake	753	Wright	Geol. Surv.
Richardson peak	3,700	Gaspé	Geol. Surv.
Richmond		Richmond	
Station	389		G. T. R.
St. Francis river, 1·8 mile W., water, 366; rail	422		G. T. R.
Rigaud	108	Vaudreuil	C. P. R.
Rigaud mountain	850	Vaudreuil	Geol. Surv.
Rimouski	54	Rimouski	I. C. R.
Riopel	35	Bonaventure	A. & L. S. R.
River Beaudette	167	Soulanges	G. T. R.
River du Chêne		Lotbinière	
Station	386		I. C. R.
River du Chêne, bed, 355; rail	382		I. C. R.
River du Loup		Temiscouata	
Station	317		I. C. R.
Temiscouata switch	325		I. C. R.
River du Loup, high water, 309; low water, 302; rail	317		Tem. R.
River du Loup, high water, 286; low water, 276; rail	315		I. C. R.
River Pierre		Portneuf	
Station	706		Q. & L. St. J. R.
River Pierre, 0·2 mile N., water, 691; rail	698		Q. & L. St. J. R.
River Rouge		Soulanges	
Station	162		G. T. R.
River Rouge, bed, 139; rail	158		G. T. R.
Rivière Ouelle		Kamouraska	
Station	48		I. C. R.
Rivière Ouelle, bed, 20; rail	43		I. C. R.
Robertson	1,205	Megantic	Q. C. R.
Roberval	352	L. St. John	Q. & L. St. J. R.
Rockfield	86	Jacques Cartier	G. T. R.
Rock Forest	701	Sherbrooke	C. P. R.
Rockhurst	327	Wright	O. N. & W. R.
Rockland	178	Labelle	C. P. R.
Rockland	404	Richmond	G. T. R.
Rosemere	91	Terrebonne	C. P. R.
Rouge, Cape, peak one mile W. of	1,955	Montmagny	Admir. chart
Rougemont		Rouville	
Station	154		Q. S. R.
Central Vermont R. R. crossing	144		Q. S. R.
Round lake	550	Argenteuil	M. & G. C. R.
Roxton Falls		Shefford	
Station	347		C. P. R.
Black river, water, 334; rail	346		C. P. R.
Royal, Mount	763	Hochelaga	City Eng.
Rush lake	1,054	Abitibi	O'Sullivan
Sabrevois	123	Iberville	Q. S. R.
Sacré Cœur	19	Rimouski	I. C. R.
Saddleback mountain	3,198	Compton	Bdy. Com.
Saddle hill	374	Saguenay	Admir. chart
Ste. Adèle	637	Terrebonne	C. P. R.
Ste. Agapit	406	Lotbinière	G. T. R.
Ste. Agathe		Terrebonne	
Station	1,194		C. P. R.
River du Nord, 1·1 mile, S., bed, 1,094; rail	1,103		C. P. R.
Summit, 1·0 mile, N	1,250		C. P. R.
Ste. Agnes		Huntingdon	
Station	195		G. T. R.
International boundary, 4·3 miles, S	171		G. T. R.
St. Aimé	90	Richelieu	Q. S. R.
St. Alban	182	Portneuf	G. N. R.
St. Alban mountain	3,170	Gaspé	Admir. chart
St. Alexandre		Kamouraska	
Station	370		I. C. R.
Summit, 1·9 mile, E	429		I. C. R.

9

QUEBEC

Locality	Elev.	County	Authority
St. Alexis....................	275	Montcalm	G. N. R.
St. Anaclet and Father Point....................	96	Temiscouata	I. C. R.
St. André	342	Kamouraska	I. C. R.
St. Andrews......	103	Argenteuil	A. & L. S. R.
Ste. Angèle.................		Rouville	
Station	157		Q. S. R.
Central Vermont R. R. crossing....	158		Q. S. R.
Summit, 0·9 mile, N........	161		Q. S. R.
Ste. Angèle de Laval, B. M., in front, north side of main entrance of R. C. church....	34·2	Nicolet	Public Work
Ste. Anne........		Kamouraska	
Station......	102		I. C. R.
Ste. Anne river, bed, 81; rail	102		I. C. R.
Ste. Anne lake....................	1,313	Gaspé	Geol. Surv.
Ste. Anne lake, "peaks on east and west sides rise to".	2,800	Gaspé	Admir. chart
Ste. Anne, Mount	2,680	Montmagny	Admir. chart
Ste. Anne de Beaupré....	17	Montmorency	Q. R. L. & P. Co.
Ste. Anne de Beaupré hills....................	2,620	Montmorency	Admir. chart
Ste. Anne de Bellevue...		Jacques Cartier	
Grand Trunk station......	122		G. T. R.
Can. Pac. station.....	113		C. P. R.
Ottawa river, above lock, high water, 77; low water, 69; rail, Can. Pac. bridge....	111		Rys. & Canals
Ste. Anne de la Pérade		Champlain	
Station	42		C. P. R.
Ste. Anne river (east bridge) water, 25; rail.......	41		C. P. R.
River St. Lawrence, flood (1865).......	26·8		Public Works
Ste. Anne des Plaines....................		Terrebonne	
Station....................	180		C. P. R.
River Ste. Anne, bed, 157; rail...........	157		C. P. R.
St. Anselme		Dorchester	
Station	506		Q. C. R.
River Etchemin, 1·1 mile, S., high water, 503; low water, 486; rail..........	524		Q. C. R.
Ste. Apollinaire................	317	Lotbinière	I. C. R.
St. Arsène		Temiscouata	
Station	274		I. C. R.
Summit, 0·7 mile, E., ground, 300; rail	292		I. C. R.
St. Augustin....................		Two Mountains	
Station	227		C. P. R.
Summit, 2·7 miles, W	275		C. P. R.
St. Augustin, St. Lawrence river, flood, (1865)....	19·5	Portneuf	Public Works
St. Barnabé....	121	St. Hyacinthe	Q. S. R.
St. Barthélemi....................		Berthier	
Can. Pac. station	36		C. P. R.
Great Northern station	178		G. N. R.
River Cachée, 1·2 mile, E., water, 142; rail.......	177		G. N. R.
St. Bazile....	202	Portneuf	C. P. R.
St. Bazile........	61	Chambly	G. T. R.
St. Boniface		St. Maurice	
Station	410		G. N. R.
East Br., Yamachiche river, 2·2 miles, W., bed, 288; rail....................	369		G. N. R.
Ste. Brigide....	165	Iberville	C. P. R.
St. Bruno....................	97	Chambly	G. T. R.
St. Canut		Two Mountains	
Station	233		G. N. R.
North mile, 1·0 mile, W., high water (1899).....	210		G. N. R.
St. Casimir....	117	Portneuf	G. N. R.
Ste. Catherine....	547	Portneuf	Q. & L. St. J. R.
Ste. Cecile....		Beauce	
Station	1,667		Q. C. R.
Summit, 4·4 miles, N., ground, 1,701; rail........	1,686		Q. C. R.
Ste. Cecile	150	Beauharnois	N. Y. C. R.
St. Charles....		Bellechasse	
Station	293		I. C. R.
Summit, 1·3 mile, N	335		I. C. R.
St. Charles, B. M. in N. W. corner of R. C. church....	44·6	St. Hyacinthe	Public Works
Ste. Christine....................	498	Portneuf	G. N. R.

QUEBEC

Locality	Elev.	County	Authority
St. Cléophas..		Berthier	
Station.......................................	470		C. P. R.
River Bayonne, 2·0 miles, S., water, 396; rail.....	400		C. P. R.
St. Clet...	177	Soulanges	C. P. R.
St. Constant.......................................		Laprairie	
Grand Trunk station...........................	95		G. T. R.
Can. Pac. station.............................	94		C. P. R.
Ste. Croix..	415	Lotbinière	I. C. R.
Ste. Cunegonde....................................	53	Hochelaga	G. T. R.
St. Cuthbert.......................................		Berthier	
Can. Pac. station.............................	41		C. P. R.
Great Northern station........................	197		G. N. R.
River Chicot, 1·2 mile, E., high water, 143; low water, 138; rail......................	193		G. N. R.
St. Cyr...	481	Richmond	G. T. R.
St. Cyrille..	280	Drummond	I. C. R.
St. Damase..	112	St. Hyacinthe	Q. S. R.
St. David...	93	Yamaska	C. P. R.
St. Denis...		St. Hyacinthe	
B. M., in N. W. corner of front of R. C. church....	53·3		Public Works
B. M., in N. W. corner of convent wall...........	48·5		Public Works
St. Dominique.....................................	159	Soulanges	G. T. R.
St. Edward..	235	Bagot	I. C. R.
Ste. Elizabeth.....................................		Joliette	
Station.......................................	177		G. N. R.
Bayonne river, high water, 161; low water, 150; rail.	177		G. N. R.
St. Eloi..	108	Temiscouata	I. C. R.
Ste. Emelie.......................................	239	Joliette	C. P. R.
Ste. Emelie, B. M., in foundation of R. C. church tower.	92·1	Lotbinière	Public Works
St. Ephrem..		Beauce	
Station.......................................	729		Q. C. R.
Le Bras river, high water, 711; low water, 705; rail.	719		Q. C. R.
St. Eugene..	273	Drummond	I. C. R.
St. Eustache......................................	98	Two Mountains	C. P. R.
St. Evariste.......................................	990	Beauce	Q. C. R.
St. Fabien..		Rimouski	
Station.......................................	445		I. C. R.
R. Sud-Ouest or Grand Bic, 1·8 mile, E., bed, 376; rail.....................................	408		I. C. R.
St. Faustin.......................................		Terrebonne	
Station.......................................	1,257		C. P. R.
Summit, 1·4 mile, S., ground, 1,426; rail.........	1,403		C. P. R.
St. Félix...	372	Joliette	C. P. R.
St. Flavie..		Matane	
Station.......................................	266		I. C. R.
Grand Metis river, 2·9 miles, E., water, 211; rail..	270		I. C. R.
Hill, six miles S. E. of......................	1,500		Admir. chart
St. Francis.......................................	557	Beauce	Q. C. R.
St. Francis, Lake (at Valleyfield)		Huntingdon	
Standard high water...........................	153·7		D. W. Com.
Standard low water............................	151·2		D. W. Com.
Extreme low water............................	149·8		D. W. Com.
Mean water...................................	152·5		D. W. Com.
St. François......................................	134	Bellechasse	I. C. R.
St. François......................................		Temiscouata	
Station.......................................	884		Tem. R.
St. Francis river, 0·9 mile, W., water, 858; rail....	886		Tem. R.
St. François du Lac...............................	81	Yamaska	A. & L. S. R.
St. François Xavier, peak 1½ miles N. W. of...........	1 970	Charlevoix	Admir. chart
St. Frederic......................................	883	Beauce	Q. C. R.
St. Gabriel.......................................		Quebec	
Station.......................................	581		Q. & L. St. J. R.
Jacques Cartier river, bed, 489; rail.............	556		Q. & L. St. J. R.
St. Gabriel.......................................		Berthier	
Station.......................................	607		C. P. R.
Summit, 0·7 mile, S...........................	629		C. P. R.
St. Gédéon..	353	L. St. John	Q. & L. St. J. R.
St. George..	203	Bagot	I. C. R.

QUEBEC

Locality	Elev.	County	Authority
St. Germain....................................		Drummond	
Can. Pac. station..........................	283		C. P. R.
Intercolonial station........................	241		I. C. R.
St. Gregoire (Mount Johnson)................	155	Iberville	Q. S. R.
St. Gregoire.................................	82	Nicolet	A. & L. S. R.
St. Guillaume................................	143	Drummond	C. P. R.
Ste. Hélène.................................	323	Kamouraska	I. C. R.
Ste. Henedine...............................		Dorchester	
Station..................................	646		Q. C. R.
Summit, 0·7 mile, E.......................	656		Q. C. R.
St. Henri...................................		Hochelaga	
Station..................................	61		G. T. R.
B. M., in front of pump-house.............	48·1		Public Works
St. Henri...................................	297	Lévis	I. C. R.
St. Henri de Mascouche......................	65	L'Assomption	C. P. R.
St. Henryville..............................	202	Lévis	Q. C. R.
St. Hermas.................................	256	Two Mountains	C. P. R.
St. Hilaire.................................		Rouville	
Station..................................	85		G. T. R.
B. M., in S. W. corner of R. C. church.....	48·6		Public Works
B. M., in east abutment of G. T. Ry. bridge.	52·8		Public Works
St. Hilaire East.............................	107	Rouville	G. T. R.
St. Honoré.................................		Temiscouata	
Station..................................	1,302		Tem. R.
Summit, 2·6 miles E......................	1,324		Tem. R.
St. Hubert..................................	90	Chambly	G. T. R.
St. Hugues.................................	113	Bagot	C. P. R.
St. Hyacinthe..............................		St. Hyacinthe	
Can. Pac. station..........................	108		C. P. R.
Grand Trunk station........................	109		G. T. R.
Que. Sou. station.........................	117		Q. S. R.
Yamaska river, water (May, 1878), 64 ; rail..	104		G. T. R.
St. Ignace, range four miles S. E. of........	1,220	Montmagny	Admir. chart
St. Isidore junction.........................	120	Laprairie	G. T. R.
St. Isidore station..........................	174	Châteauguay	G. T. R.
St. Jacques.................................	204	Montcalm	G. N. R.
St. Janvier.................................	217	Terrebonne	C. P. R.
St. Jean Chrysostome.......................		Lévis	
Station..................................	214		I. C. R.
River Etchemin, 2·5 miles E., water, 203 ; rail.	252		I. C. R.
St. Jean des Chaillons.......................		Lotbinière	
Station..................................	145		L. & M. R.
B. M., in N. W. corner of R. C. church, 4 feet above ground..........................	153·3		Public Works
Levasseur's wharf, floor level..............	15·3		Public Works
River St. Lawrence—			
Standard low water.......................	1·6		Public Works
Standard high water......................	21·0		Public Works
Extreme high water......................	26·8		Public Works
St. Jean de Dieu P. O.......................	505	Temiscouata	I. C. R.
St. Jean Port Joli..........................		L'Islet	
Station..................................	176		I. C. R.
River St. Jean, 1·9 mile, W., bed, 140 ; rail..	157		I. C. R.
Summit, 1·2 mile, E.......................	185		I. C. R.
Peak 6½ miles E. of.......................	1,240		Admir. chart
St. Jérôme.................................		L. St. John	
Station..................................	357		Q. & L. St. J. R.
Summit, 1·6 mile, W......................	425		Q. & L. St. J. R.
St. Jérôme.................................		Terrebonne	
Can. Pac. station..........................	308		C. P. R.
Great Nor. station........................	292		G. N. R.
Great Northern Railway crossing, 1·5 mile, S.	262		C. P. R.
St. Jérôme junction.........................	263	Terrebonne	G. N. R.
St. Joachim.................................	26	Montmorency	Q. R. L. & P. Co.
St. Joachim hill.............................	610	Montmagny	Admir. chart
St. Johns...................................		St. Johns	
Can. Pac. station..........................	116		C. P. R.
Grand Trunk station........................	122		G. T. R.
Can. Pac. Ry. crossing......................	115		G. T. R.

QUEBEC

Locality	Elev.	County	Authority
St. Johns		St. Johns	
Richelieu river—			
Standard low water	92·4		Public Works
Lowest normal water..........	92·0		Public Works
Extreme low water	90·4		Public Works
Mean water (1871 to 1895).........	94·5		Public Works
Extreme high water, one-half mile above lock (April, 1869)	99·0		Public Works
Bench marks—			
N. E. corner of G. Nolin's house, opposite lock..	105·8		Public Works
Coping, upper end of lock............	100·4		Public Works
Front of Montgomery's house, corner of Lemoine and Champlain streets.............	116·3		Public Works
Top of cap of bench well, inside barracks........	101·3		Public Works
St. John, Lake, high water, 341 ; low water....	314	L. St. John	Q. & L. St. J. R.
St. John, Mount................	1,416	Saguenay	Admir. chart
St. Joseph.................	490	Beauce	Q. C. R.
St. Joseph de Lévis..................		Lévis	
Station	162		I. C. R.
B. M., in abutment of I. C. Ry. crossing, opposite graving dock	57·1		Public Works
B. M., in second altar step, S. W. end of graving dock	14·9		Public Works
St. Joseph de Sorel, B. M., in N. W. corner of Messrs. McCarthy's house, 3 feet above ground	32·9	Richelieu	Public Works
St. Joseph mountain	1,000	Gaspé	Admir. chart
St. Jovite..................		Terrebonne	
Station	704		C. P. R.
Devil river, 2·1 miles, W., high water, 660 ; low water, 648 ; rail................	686		C. P. R.
St. Jude		St. Hyacinthe	
Station	100		Q. S. R.
River Salvaille, water, 44 ; rail............	100		Q. S. R.
Ste. Julie.................	475	Megantic	G. T. R.
Ste. Julienne.................	273	Montcalm	G. N. R.
St. Justin	250	Maskinongé	G. N. R.
Ste. Justine................	244	Vaudreuil	C. A. R.
St. Lambert.................	74	Chambly	G. T. R.
St. Laurent	134	Jacques Cartier	G. T. R.
St. Lazare	162	Vaudreuil	C. P. R.
St. Leonard................		Nicolet	
Station	237		I. C. R.
Junction	232		I. C. R.
North-east branch Nicolet river, water, 149 ; bed, 144 ; rail................	226		I. C. R.
St. Liboire (summit)................	289	Bagot	G. T. R.
St. Lin................	238	L'Assomption	G. N. R.
St. Lin................		Terrebonne	
Station	206		C. P. R.
River L'Achigan, bed, 158 ; rail....	197		C. P. R.
St. Lin junction................	219	Terrebonne	C. P. R.
St. Louis		Beauharnois	
Station	136		C. A. R.
St. Louis river, water, 130 ; rail......	136		C. A. R.
B.M., in east abutment of C. A. Ry. bridge over St. Louis river.	131·6		D. W. Com.
St. Louis................	103	Richelieu	Q. S. R.
St. Louis de Ha-Ha................	825	Temiscouata	Tem. R.
St. Louis mountain................	1,000	Bonaventure	Admir. chart
St. Louis, Lake................		Beauharnois	
At Lachine, extreme high water	73·7		D.W. Com.
At Lachine, extreme low water	64·6		D.W. Com.
At Melocheville, high water.........	75·2		D.W. Com.
At Melocheville, extreme low water.....	66·4		D.W. Com.
At Cascades Point, standard low water	67·7		D.W. Com.
Ste. Louise................	119	L'Islet	I. C. R.
St. Luc junction................	158	Hochelaga	C. P. R.
Ste. Luce................	150	Rimouski	I. C. R.

QUEBEC

Locality	Elev.	County	Authority
Ste. Madeleine............		St. Hyacinthe	
Grand Trunk station	117		G. T. R.
Que. Sou. station	113		Q. S. R.
Huron river, 0·7 mile, W., water, 87 ; rail	103		G. T. R.
St. Malo........		Compton	
Station	1,501		Me. C. R.
Summit, 1·3 mile, S.........	1,585		Me. C. R.
Ste. Margaret....		Terrebonne	
Station	900		C. P. R.
Indian Mountain summit, 1·5 mile, N., ground, 1,073 ; rail .	1,025		C. P. R.
St. Martin station.... ..	93	Laval	C. P. R.
St. Martin junction........	110	Laval	C. P. R.
Ste. Martine...........		Châteauguay	
Station	128		G. T. R.
Châteauguay river, bed, 96 ; rail..	126		G. T. R.
St. Mary	479	Beauce	Q. C. R.
St. Maurice......		Champlain	
Station	184		C. P. R.
River Champlain, 1·0 mile, S., water, 135 ; rail....	158		C. P. R.
St. Michel	176	Bellechasse	I. C. R.
St. Michel.............	188	Napierville	G. T. R.
Ste. Modeste...........	545	Temiscouata	Tem. R.
St. Moise............		Matane	
Station	640		I. C. R.
Summit, 2·5 miles, S........	751		I. C. R.
St. Monique	162	Nicolet	I. C. R.
St. Narcisse..	384	Champlain	C. P. R.
St. Nicholas........		Lévis	
Station	255		I. C. R.
Terrebonne river, 1·1 mile, E., bed, 229 ; rail ...	248		I. C. R.
Baker's wharf, floor level.	12·9		Public Works
St. Norbert		Berthier	
Station	181		G. N. R.
Bonaventure river, water, 144 ; rail., ...	177		G. N. R.
St. Octave	346	Matane	I. C. R.
St. Ours, B. M., in south side of R. C. church.	52·5	Richelieu	Public Works
St. Paschal		Kamouraska	
Station	195		I. C. R.
Kamouraska river, 0·9 mile, W., bed, 164 ; rail	179		I. C. R.
St. Paulin		Maskinongé	
Station	537		G. N. R.
Summit, 0·6 mile, E......	551		G. N. R.
Ste. Perpétue.....	228	Nicolet	I. C. R.
St. Peter, Lake			
Standard low water	11·1		Public Works
Standard high water..	30·0		Public Works
At Pointe du Lac, flood, (1865)...... ..	32·5		Public Works
St. Philippe de Laprairie...	117	Laprairie	C. P. R.
St. Philippe de Néri. /	145	Kamouraska	I. C. R.
St. Philippe West		Argenteuil	
Can. Pac. station	263		C. P. R.
Great Nor. station.	247		G. N. R.
Summit, 2·8 miles, W, ground, 341 ; rail	335		C. P. R.
St. Philomène........	140	Châteauguay	G. T. R.
Ste. Philomène.	208	Lotbinière	L. & M. R.
St. Pie.......	131	Bagot	C. P. R.
St. Pierre...........	131	Bellechasse	I. C. R.
St. Pierre les Becquets, B. M., in N. W. corner of front of R. C. church.	103·0	Nicolet	Public Works
St. Polycarpe station	176	Soulanges	C. A. R.
St. Polycarpe junction.............	191	Soulanges	C. P. R.
St. Prime..............	124	Drummond	C. P. R.
St. Prosper	281	Champlain	G. N. R.
St. Raymond........		Portneuf	
Station	452		Q. & L. St. J. R.
River Ste. Anne, high water, 441 ; low water, 428 ; rail......	446		Q. & L. St. J. R.
Summit, 2·6 miles, W........	562		Q. & L. St. J. R.

QUEBEC

Locality	Elev.	County	Authority
St. Rémi	196	Napierville	G. T. R.
St. Robert		Richelieu	
Que. Sou. station	71		Q. S. R.
Atl. & L. S. station	42		A. & L. S. R.
St. Roch	55	Richelieu	N. S. R.
St. Roch des Aulnaies, peak seven miles S. E. of	1,666	L'Islet	Admir. chart
St. Roch Nord	17	Quebec	Q. R. L. & P. Co.
St. Romuald	59	Lévis	I. C. R.
Ste. Rosalie		Bagot	
Can. Pac. station	107		C. P. R.
Grand Trunk station	114		G. T. R.
Can. Pac. Ry. crossing	111		G. T. R.
Ste. Rose		Laval	
Station	85		C. P. R.
River des Mille Iles, water, 68; rail	85		C. P. R.
Ste. Rose	504	Temiscouata	Tem. R.
St. Sabine	193	Missisquoi	C. P. R.
St. Scholastique	238	Two Mountains	C. P. R.
St. Simon	118	Bagot	C. P. R.
St. Simon	206	Rimouski	I. C. R.
Ste. Sophie	251	Terrebonne	G. N. R.
St. Stanislas	157	Beauharnois	N. Y. C. R.
St. Stanislas	394	Champlain	G. N. R.
St. Telesphore	212	Soulanges	C. P. R.
St. Thècle	510	Champlain	G. N. R.
St. Thérèse	120	Terrebonne	C. P. R.
St. Timothée	174	Beauharnois	N. Y. C. R.
St. Tite		Champlain	
Station	465		G. N. R.
River des Envies, bed, 421; rail	467		G. N. R.
Summit, 2.3 miles, E.	543		G. N. R.
St. Tite junction	504	Champlain	G. N. R.
Ste. Ursule		Maskinongé	
Station	360		G. N. R.
Maskinongé river, bed, 137; rail	308		G. N. R.
St. Valentin		St. Johns	
B. M., in foundation of Boileau's hotel	105.4		Public Works
Top of cap of bench well, opp. Bissonnette's house	96.6		Public Works
St. Valier	156	Bellechasse	I. C. R.
St. Victor de Tring	722	Beauce	Q. C. R.
St. Vincent de Paul	75	Laval	C. P. R.
St. Wenceslas	275	Nicolet	I. C. R.
St. Zotique	159	Soulanges	G. T. R.
Salmon Lake station	503	Matane	I. C. R.
Sam or Mercier lake, ice	740	Labelle	C. P. R.
Sand lake	575	Abitibi	O'Sullivan
Sandgirt lake	1,650	Ashuanipi	Geol. Surv.
Sandy Bay siding	699	Matane	I. C. R.
Sandy-beach lake	1,222	Champlain	O'Sullivan.
Sault aux Cochons, peak 1½ miles W. of	2,365	Montmagny	Admir. chart
Sault aux Récollets		Jacques Cartier	
Station	75		C. P. R.
River des Prairies, water, 57; rail	75		C. P. R.
Savage Mills	567	Shefford	C. P. R.
Sawyerville	884	Compton	Me. C. R.
Sayabec	578	Matane	I. C. R.
Scaumenac, Mount	1,745	Bonaventure	Admir. chart
Scotstown		Compton	
Station	1,193		C. P. R.
Salmon river, water, 1,173; rail	1,191		C. P. R.
Scott	475	Dorchester	Q. C. R.
Seven-league lake, high water, 541; low water	522	Pontiac	Rys & Canals
Shabogama lake	850	Abitibi	O'Sullivan
Shagois lake	1,202	Champlain	O'Sullivan
Shawbridge	599	Terrebonne	C. P. R.
Shawenegan Junction		Champlain	
Station	460		G. N. R.
Shawenegan river, 1.1 mile, W., water, 381; rail	463		G. N. R.
Shawenegan Falls station	2	Champlain	G. N. R.

QUEBEC

Locality	Elev.	County	Authority
Shawville....		Pontiac	
Station	527		P. P. J. R.
Summit, 1·0 mile, W.	610		P. P. J. R.
Sherbrooke.		Sherbrooke	
Grand Trunk station	485		G. T. R.
Can. Pac. station.	609		C. P. R.
City hall, floor.	521		City Eng.
Pond, water.	583		City Eng.
R. C. hospital, main floor	707		City Eng.
Reservoir, surface when full	711		City Eng.
Waterworks pump-house, floor.	570		City Eng.
St. Francis river, high water, 482 ; low water	461		City Eng.
Magog river, water (June, 1878), 462 ; rail	487		G. T. R.
Sherrington	184	Napierville	G. T. R.
Shickshock mountains	3,500	Gaspé	Geol. Surv.
Simon lake.	915	Abitibi	O'Sullivan
Simon, Lac.	581	Portneuf	Q. & L. St. J. R.
Simon, Lake.	702	Labelle	Geol. Surv.
Single-tree mountain	2,060	Montmorency	Admir. chart
Sixteen-island lake	861	Argenteuil	M. & G. C. R.
Skunk lake	1,482	Champlain	Geol. Surv.
Smith Mills	613	Stanstead	B. & M. R.
Snake Creek station	543	Pontiac	C. P. R.
Somerset (now Plessisville)	441	Megantic	G. T. R.
Sorcerer mountain	1,575	Abitibi	O'Sullivan
Sorel		Richelieu	
Can. Pac. station.	47		C. P. R.
Que. Sou. station.	49		Q. S. R.
B. M., top of boundary stone on S.E. Ry. property.	46·5		Public Works
B. M., in foundation of market, 3½ feet above ground.	39·7		Public Works
R. & O. Co.'s wharf, floor level	+26·8		Public Works
Zero of Montreal Harbour Commissioners' gauge..	— 5·6		Public Works
River St. Lawrence—			
Standard low water.	12·1		Public Works
Standard high water.	27·4		Public Works
Extreme high water.	31·2		Public Works
Flood (1865).	30·9		Public Works
Sorel West.	30	Richelieu	S. S. R.
Soskumika lake	600	Abitibi	O'Sullivan
Soulards	308	Nicolet	I. C. R.
South Bolton	711	Brome	M. & B. R. V. R.
South Durham		Drummond	
Station	608		G. T. R.
Summit, 0·5 mile, W.	612		G. T. R.
South mountain	2,413	Gaspé	Geol. Surv.
South Roxton	515	Shefford	C. P. R.
South Stukely.	837	Shefford	C. P. R.
Spring Hill		Compton	
Station	1,690		C. P. R.
Summit, 1·9 mile, W.	1,698		C. P. R.
Spruce	1,466	Compton	C. P. R.
Stadacona		Quebec	
Station	1,204		Q. & L. St. J. R.
Rognons river, bed, 1,193 ; rail.	1,202		Q. & L. St. J. R.
Summit, 3·4 miles, N., ground, 1,253; rail	1,245		Q. & L. St. J. R.
Stanbridge	167	Missisquoi	C. P. R.
Stanfold	525	Arthabaska	G. T. R.
Stanstead junction	748	Stanstead	B. & M. R.
Stearns	1,619	L. Megantic	C. P. R.
Stoke mountains	1,600-1,800	Richmond	Geol. Surv.
Stottsville	151	St. Johns	G. T. R.
Strathmore	89	Jacques Cartier	G. T. R.
Summit	411	Portneuf	C. P. R.
Summit lake.	761	Labelle	C. P. R.
Sutton		Brome	
Station	582		C. P. R.
Summit, 1·6 mile, N.	644		C. P. R.
Sutton junction		Brome	
Station	553		C. P. R.
Drummondville branch switch	559		C. P. R.

QUEBEC

Locality	Elev.	County	Authority
Sutton mountains	3,000	Brome	Geol. Surv.
Sweetsburg	441	Missisquoi	C. P. R.
"T" lake, high water. 856 ; low water	849	Pontiac	C. P. R.
Table Roulante or Percée mountain	1,230	Gaspé	Admir. chart
Table-top mountain	4,000	Gaspé	Geol. Surv.
Tadoussac hill	1,100	Saguenay	Admir. chart
Talbot		Portneuf	
Station	970		Q. & L. St. J. R.
Summit, 0·8 mile, N	978		Q. & L. St. J. R.
Terrace mountain	1,957	Gaspé	Geol. Surv.
Terrebonne		L'Assomption	
Station	63		C. P. R.
Mille Iles river, water, 30 ; rail	58		C. P. R.
Thanvette	108	Vaudreuil	C. P. R.
Thetford (now Kingsville)		Megantic	
Station	1,026		Q. C. R.
Thetford bridge, 1·8 mile, E., bed, 1,043 ; rail	1,063		Q. C. R.
Thompson or McConnell lake	874	Pontiac	Geol. Surv.
Three Mountains, Lake of	788	Labelle	Geol. Surv.
Three Rivers		St. Maurice	
Station	56		C. P. R.
River St. Maurice, water, 22 ; rail	67		C. P. R.
Zero of gauge	9·3		Public Works
River St. Lawrence—			
Standard low water	9·5		Public Works
Standard high water	25·3		Public Works
Highest known flood	29·6		Public Works
Thurso		Labelle	
Station	186		C. P. R.
Blanche river, 2·2 miles, W., water, 128; rail	164		C. P. R.
Timiskaming lake, high water, 592 ; low water	578	Pontiac	Rys. & Canals
Timiskaming	593	Pontiac	C. P. R.
Titus	427	Richmond	G. T. R.
Tongue lake	857	Abitibi	O'Sullivan
Touradif lake, Upper	573	Rimouski	I. C. R.
Touradif lake, Lower	563	Rimouski	I. C. R.
Tourmente, Cape	1,874	Montmagny	Admir. chart
Tourmente, Cape, chapel	1,692	Montmagny	Admir. chart
Tourmente, Cape, hill three miles N.W. of	1,745	Montmagny	Admir. chart
Tracadigash mountain	1,230	Bonaventure	Admir. chart
Travers, Lac de	1,212	Champlain	O'Sullivan
Trembling lake	761	Labelle	Geol. Surv.
Trembling mountain	2,474	Labelle	Geol. Surv.
Tring station	1,135	Beauce	Q. C. R.
Tring and Megantic junction	1,050	Beauce	Q. C. R.
Triton Club		Quebec	
Station	1,172		Q. & L. St. J. R.
North branch Batiscan river, high water, 1,146 ; low water, 1,139 ; rail	1,165		Q. & L. St. J. R.
' Beaver Pond ' summit, 1·0 mile, S	1,214		Q. & L. St. J. R.
Trois Pistoles	112	Temiscouata	I. C. R.
Trois Saumons		L'Islet	
Station	100		I. C. R.
Peak ten miles E. of	2,090		Admir. chart
Trout lake	1,175	L. St. John	O'Sullivan
Trudel	1,597	Compton	C. P. R.
Turcot	63	Hochelaga	C. P. R.
Tush lake	841	Abitibi	O'Sullivan
Two Mountains, Lake of, high water, 77 ; low water	69	Two Mountains	Rys. & Canals
Upton	202	Bagot	G. T. R.
Valcartier		Quebec	
Station	556		Q. & L. St. J. R.
Summit, 1·6 mile, W	581		Q. & L. St. J. R.
Valcourt	727	Shefford	O. M. R.
Valleyfield		Beauharnois	
Can. Atl. station	159		C. A. R.
N. Y. Cent. station	159		N. Y. C. R.
Upper lock, sill	140·3		D. W. Com.

QUEBEC

Locality	Elev.	County	Authority
Valleyfield..		Beauharnois	
Lake St. Francis—			
Standard high water (May, 1870)...............	153·7		D. W. Com.
Standard low water.......................	151·2		D. W. Com.
Extreme low water (Nov. 9, 1895)...........	149·8		D. W. Com.
Mean water..............................	152·5		D. W. Com.
Val Morin..	1,025	Terrebonne	C. P. R.
Valois...		Jacques Cartier	
Can. Pac. station..........................	93		C. P. R.
Grand Trunk station.......................	89		G. T. R.
Varennes..		Verchères	
Station...	49		S. S. R.
B.M., in south side of R.C. church............	53·3		Public Works
River St. Lawrence, standard low water........	16·3		Public Works
R. & O. Co's wharf, floor level...............	27·8		Public Works
Highest known flood, half mile above Varennes wharf....	38·2		Public Works
Vaucluse..		L'Assomption	
Station...	79		C. P. R.
L'Assomption river, 1·1 mile, W., bed, 19 ; rail....	66		C. P. R.
Vaudreuil..		Vaudreuil	
Can. Pac. station..........................	87		C. P. R.
Grand Trunk station.......................	85		G. T. R.
Ottawa river, Grand Trunk bridge, rail.......	92		G. T. R.
Ottawa river, Can. Pac. bridge, rail..........	94		C. P. R.
Venosta...	534	Wright	O. N. & W. R.
Verchères...		Verchères	
Station...	63		S. S. R.
B.M., in pillar, N.W. side of main entrance of R.C. church..	63·4		Public Works
R. & O. Co's wharf, floor level..............	26·9		Public Works
River St. Lawrence—			
Standard low water......................	14·8		Public Works
Standard high water.....................	29·6		Public Works
Extreme high water......................	36·1		Public Works
Versailles...		Iberville	
Station...	176		C. P. R.
Summit, 1·3 mile, W......................	195		C. P. R.
Victoria, Grand lake..............................	960	Pontiac	O'Sullivan
Vinton...	367	Pontiac	P. P. J. R.
Wagan lake...	568	Rimouski	I. C. R.
Wakefield..		Wright	
Station...	326		O. N. & W. R.
Peche river, water, 306 ; rail...............	328		O. N. & W. R.
Wakefield lake.....................................	531	Labelle	Geol. Surv.
Wakonichi hill......................................	1,440	Mistassini	O'Sullivan
Wakonichi lake.....................................	1,240	Mistassini	O'Sullivan
Waltham..		Pontiac	
Station...	367		P. P. J. R.
Ottawa river, high water, 353 ; low water....	340		P. P. J. R.
Warden...	674	Shefford	C. P. R.
Warwick..	480	Arthabaska	G. T. R.
Waswanipi lake.....................................	680	Abitibi	O'Sullivan
Waterloo...		Shefford	
Station...	677		C. P. R.
Central Vermont R. R. crossing..............	710		C. P. R.
North branch Yamaska river, bed, 668 ; rail....	675		C. P. R.
Water-swamp lake..................................	1,395	Champlain	Geol. Surv.
Waterville..		Compton	
Station...	645		G. T. R.
Coaticook river, water, 619 ; rail..........	644		G. T. R.
Weedon...		Wolfe	
Station...	815		Q. C. R.
Summit, 1·6 mile, W.......................	882		Q. C. R.
Weedon tank..	814	Wolfe	Q. C. R.
West Brome...		Brome	
Station...	412		C. P. R.
South branch Yamaska river, 0·8 mile, E., low water, 409 ; rail..................................	430		C. P. R.

QUEBEC

Locality	Elev.	County	Authority
Westbury station...............................	659	Compton	Q. C. R.
Westbury tank....	647	Compton	Q. C. R.
Western junction.	138	Jacques Cartier	C. P. R.
Westmount	152	Hochelaga	C. P. R.
West point........	400	Bonaventure	Admir. chart
West Shefford.		Shefford	
Station.........	430		C. P. R.
North branch Yamaska river, 1·2 mile, E., water, 403; rail	436		C. P. R.
Wetetnagami lake......	1,055	Abitibi	O'Sullivan
Wettigo lake	599	Mistassini	O'Sullivan
White station...........	181	Huntingdon	G. T. R.
White lake........	872	Pontiac	Geol. Surv.
White Bear lake....	1,450	Champlain	Geol. Surv.
White Beaver lake	883	Pontiac	Geol. Surv.
Whitefish lake..	1,190	L. St. John	O'Sullivan
Wickham.......		Drummond	
Station	378		C. P. R.
Black river, 2·0 miles W., high water, 283; rail,...	302		C. P. R.
Summit, 3·2 miles S...................	456		C. P. R.
Willow....	130	Jacques Cartier	G. T. R.
Windsor......	419	Richmond	G. T. R.
Winokapau lake	750	Ashuanipi	Geol. Surv.
Windy lake.............	1,040	Abitibi	O'Sullivan
Wissick, Mount	1,033	Temiscouata	Bdy. Com.
Woodlands.....	103	Châteauguay	N. Y. C. R.
Wyman.	394	Pontiac	P. P. J. R.
Yamachiche........		St. Maurice	
Station	37		C. P. R.
Yamachiche river, 1·5 mile, E., water, 33; rail....	46		C. P. R.
Yamaska		Yamaska	
Station....	53		Q. S. R.
B.M., S.E. corner, front of R. C. church ....:.....	57·5		Public Works
Yamaska East	53	Yamaska	Q. S. R.

SASKATCHEWAN

Locality	Elev.	Lat.		Long.		Authority
		°	′	°	′	
Athapapuskow lake........................	935	54	35	101	40	Geol. Surv.
Beaver lake.........	1,075	54	30	102	15	Geol. Surv.
Blackfoot hills......	2,425	53	14	110	12	Geol. Surv.
Cedar lake....	828	53	15	100	00	Geol. Surv
Clark crossing..	1,630	52	14	106	36	C. P. R.
Clearwater lake..........	860	54	00	101	00	Geol. Surv.
Clouston	1,481	53	07	105	51	C. P. R
Cranberry lake, Lower.	934	54	10	101	10	Geol. Surv.
Cranberry lake, Upper	935	54	05	101	15	Geol. Surv.
Cross lake (Saskatchewan river)................	822	53	10	99	35	Geol. Surv.
Cross lake (Nelson river)...	665	54	40	97	40	Geol. Surv.
Cumberland lake....	870	54	00	102	15	Geol. Surv.
Cygnet lake, high water................	1,786	52	05	103	55	C. P. R.
Duck lake station.................	1,645	52	49	106	13	C. P. R.
Erwood.	1,078	52	51	102	13	C. N. R.
File lake	920	54	55	100	20	Geol. Surv.
Goose lake...	905	54	25	101	30	Geol. Surv.
Gooseberry lake..	2,256	52	07	110	15	Geol. Surv.
Great Playgreen lake...	710	54	15	98	10	C. P. R.
Grindlay..	1,662	52	01	106	35	C. P. R.
Gull lake..	420	56	20	95	20	Geol. Surv.
Hague	1,671	52	29	106	25	Geol. Surv.
Little Playgreen lake....	710	54	00	98	00	C. P. R.
Macdowall........	1,540	53	00	106	00	C. P. R.
Moose lake....	846	54	45	100	15	Geol. Surv.
Net-setting lake...............	685	55	00	98	40	Geol. Surv.
Neutral Hills, 'The Nose'......	2,995	52	10	111	10	Geol. Surv.
Osler.	1,673	52	21	106	32	C. P. R.
Pasquia hills	2,000	53	00	103	00	Geol. Surv.
Pine Island or Cumberland lake	870	54	00	102	15	Geol. Surv.
Pipestone lake................	665	54	30	97	35	Geol. Surv.
Prince Albert..	1,398	53	12	105	47	C. P. R.
Red Deer lake..	875	52	55	101	20	Geol. Surv.
Reed lake......	880	54	35	100	25	Geol. Surv.
Roddick	1,607	52	55	106	08	C. P. R.
Rosthern	1,657	52	40	106	19	C. P. R.
Sandy lake.........	700	54	50	98	50	Geol. Surv.
Saskatoon.....	1,574	52	08	106	39	C. P. R.
Sipiwesk lake..............	565	55	00	97	30	Geol. Surv.
Split lake	440	56	10	96	15	Geol. Surv.
Sturgeon lake	870	54	10	101	55	Geol. Surv.
Sounding lake...........	2,165	52	10	110	30	Geol. Surv.
Sweet Herb or Wekusko lake....	800	54	45	99	50	Geol. Surv.
Thunder hill........................	1,997	52	00	109	38	C. P. R.
Wekusko or Sweet Herb lake. ...; ...., ..	800	54	45	99	50	Geol. Surv.
Winnipeg lake, high water, 713; low water, 708; mean water...	710	53	00	98	00	C. P. R.

UNGAVA

Locality	Elev.	Lat.	Long.	Authority
		° ′	° ′	
Allan, Mount	1,800	50 49	80 02	Admir. chart
Altagaiyaivik or Monkey hill	2,170	55 00	59 14	Admir. chart
Aulatsivik or Newark island	2,733	56 55	61 15	Admir. chart
Bache, Mount	2,150	59 50	64 05	Admir. chart
Big island (Kikkertaksoak)	700	58 33	62 42	Admir. chart
Bishops Mitre	3,000	57 50	61 55	Admir. chart
Blow-me-down, Mount	3,000	58 48	63 05	Admir. chart
Bluff cape	719	52 50	55 49	Admir. chart
Button island	1,500	60 45	64 30	Admir. chart
Cape St. Michael hills	895	52 49	55 49	Admir. chart
Cape Wolstenholme range	2,000	62 30	77 00	Admir. chart
Chidley, Cape, hill behind	1,500	60 30	64 30	Admir. chart
Clearwater lake	790	56 15	75 15	Geol. Surv.
Finger hill	2,500	58 00	62 10	Admir. chart
Ford Harbour hills	1,000	56 28	61 24	Admir. chart
George island	750	54 15	57 20	Admir. chart
Granby island	461	52 33	55 43	Admir. chart
Harrison, Cape	1,065	54 55	57 56	Admir. chart
Hawke island	825	53 05	55 46	Admir. chart
Itomami, Lake	1,940	53 00	69 20	Geol. Surv.
Jacopie, Lake	1,630	53 40	64 30	Geol. Surv.
Kaipokok hill	895	55 09	59 26	Admir. chart
Kangardlirasuk peaks	5,000–6,000	59 35	63 50	Admir. chart
Kaniapiskau, Lake	1,850	53 50	69 20	Geol. Surv.
Kiglapait mountain	2,000	57 07	61 25	Admir. chart
Kikkertaksoak (Big island)	700	58 33	62 42	Admir. chart
Kikkertaksoak or Spracklins island	465	56 08	60 45	Admir. chart
Mansfield island	300	62 00	80 00	Admir. chart
Mealy mountains	1,482	53 42	57 15	Admir. chart
Menihek lakes	1,750	54 00	66 40	Geol. Surv.
Michikamau, Lake	1,650	54 00	64 00	Geol. Surv.
Monkey hill (Altagaiyaivik)	2,170	55 00	59 14	Admir. chart
Nachvak mountain	3,400	59 00	63 50	Admir. chart
Nachvak range	5,000–6,000	59 05	64 00	Admir. chart
Nanuktut or White Bear island	1,600	57 55	61 40	Admir. chart
Nanuktok (White Bear cape)	1,000	59 20	63 25	Admir. chart
Natuakami lake	520	57 13	72 20	Geol. Surv.
Newark (Aulatsivik) island	2,733	56 55	61 15	Admir. chart
Nichikun, Lake	1,760	53 05	70 45	Geol. Surv.
Partridge Head	551	56 10	55 45	Admir. chart
Petitsikapau lake	1,650	53 50	65 30	Geol. Surv.
Pownal or Paul island	1,195	56 32	61 32	Admir. chart
Ragged island	640	55 00	58 13	Admir. chart
Razorback, Mount	3,000	59 12	63 25	Admir. chart
St. Charles, Cape	654	52 13	55 38	Admir. chart
Sandgirt lake	1,675	54 40	66 30	Geol. Surv.
Seal lake, Lower	860	56 30	74 30	Geol. Surv.
Ship Harbour island	721	52 39	55 45	Admir. chart
Sophia Harbour hill	584	52 25	55 43	Admir. chart
Spracklins (Kikkertaksoak) island	465	56 08	60 45	Admir. chart
Square island	497	52 46	55 51	Admir. chart
Stony island	670	53 00	55 47	Admir. chart
Table hill	2,000	57 45	61 40	Admir. chart
'The Domes'	2,000	58 27	62 40	Admir. chart
Tunungayualuk island	800	56 08	61 06	Admir. chart
Uivuk	1,000	58 30	62 35	Admir. chart
Watchman island	700	58 15	62 05	Admir. chart
White Bear cape (Nanuktok)	1,000	59 20	63 25	Admir. chart
White Bear island (Nanuktut)	1,600	57 55	61 40	Admir. chart
Winokapau, Lake	750	53 10	63 00	Geol. Surv.

YUKON

Locality	Elev.	Lat.	Long.	Authority
		° ′	° ′	
Atlin lake	2,200	59 45	133 45	Geol. Surv.
Atlin lake, Little	2,270	60 15	133 55	Geol. Surv.
*Augusta, Mount	14,900	60 18	140 28	Bdy. Com., Can.
Bennett lake	2,161	60 05	134 45	W. P. & Y. R.
Billings, Mount	7,950	61 15	128 50	Geol. Surv.
Campbell mountains	5,500	61 20	129 40	Geol. Surv.
Caribou	2,171	60 11	134 42	W. P. & Y. R.
Chilkat pass	4,950	60 09	136 04	Krause
*Cook, Mount	13,700	60 10	139 59	Bdy. Com., Can.
Cowley	2,527	60 32	134 54	W. P. & Y. R.
Dawson	1,200	64 02	139 25	Geol. Surv.
Dawson range	7,000	62 25	137 00	Geol. Surv.
Dugdale	2,437	60 35	134 56	W. P. & Y. R.
Finlayson lake	3,105	61 42	130 30	Geol. Surv.
Fort Selkirk	1,555	62 47	137 17	Geol. Surv.
Frances lake	2,577	61 20	129 30	Geol. Surv.
Glenlyon mountains	5,300	62 30	134 15	Geol. Surv.
Golden Horn mountain	5,610	60 33	134 58	Geol. Surv.
Haeckel hill	3,900	60 47	135 18	Geol. Surv.
Hayes peak	5,050	60 18	133 12	Geol. Surv.
*Hubbard, Mount	16,400	60 21	139 02	Bdy. Com., Can.
Hutchi village	2,600	61 24	136 30	Interior.
Ingram, Mount	8,000	60 45	135 39	Geol. Surv.
Jubilee mountain	6,380	60 05	134 04	Geol. Surv.
Kusawa lake	2,565	60 30	135 45	Geol. Surv.
Laberge, Lake	2,100	61 15	135 05	Geol. Surv.
Lansdowne, Mount	6,140	60 21	134 28	Geol. Surv.
Lansdowne	2,292	60 16	134 46	W. P. & Y. R.
Laurier, Mount	5,265	60 05	134 48	Geol. Surv.
Logan, Mount	19,539	60 35	140 21	Russell
Logan peak	9,000	61 33	129 04	Geol. Surv.
Lorne	2,362	60 20	134 48	W. P. & Y. R.
Lorne, Mount	6,400	60 27	134 38	Geol. Surv.
Macmillan mountains	4,000	62 55	135 30	Geol. Surv.
Marsh, Lake	2,160	60 30	134 15	Geol. Surv.
M'Clintock peak	6,000	60 48	133 47	Geol. Surv.
Michie mountain	5,540	60 30	134 01	Geol. Surv.
Miners range	4,500	61 00	135 15	Geol. Surv.
Newton, Mount	13,860	60 20	140 53	Bdy. Com. Can.
Pelly and Liard portage, west end	2,965	60 47	131 02	Geol. Surv.
Pelly and Liard portage, summit	3,150	61 44	130 40	Geol. Surv.
Pelly mountains	7,000	62 40	132 00	Geol. Surv.
Robinson	2,502	60 27	134 51	W. P. & Y. R.
*Seattle, Mount	10,000	60 05	139 12	Bdy. Com., Can.
Semenof hills	4,000	61 55	134 50	Geol. Surv.
Sifton mountain	7,500	61 15	136 00	Geol. Surv.
Simpson mountains	6,500	60 45	129 30	Geol. Surv.
Simpson Tower	5,230	61 25	129 22	Geol. Surv.
Tagish lake	2,161	60 10	134 15	Geol. Surv.
Tent peak	7,860	61 51	128 40	Geol. Surv.
Teslin lake	2,400	60 15	132 45	Geol. Surv.
'Three Guardsmen' mountains	7,300	61 24	137 15	Geol. Surv.
*Vancouver, Mount	15,617	60 21	139 42	Bdy. Com., Can.
Wette-Lea	2,452	60 24	134 50	W. P. & Y. R.
Whitehorse	2,090	60 44	135 04	W. P. & Y. R.
White, Mount	5,000	60 18	133 54	Geol. Surv.
Wigan	2,372	60 39	135 02	W. P. & Y. R.